bees

NATURE'S
LITTLE WONDERS

CANDACE SAVAGE

bees

 David Suzuki Foundation

GREYSTONE BOOKS
D&M PUBLISHERS INC.
Vancouver/Toronto/Berkeley

Greystone Books
An imprint of D&M Publishers Inc.
2323 Quebec Street, Suite 201
Vancouver BC Canada V5T 4S7
www.greystonebooks.com

David Suzuki Foundation
219–2211 West 4th Avenue
Vancouver BC Canada V6K 4S2

Cataloguing data available from Library and Archives Canada
ISBN 978-1-55365-321-9 (cloth)
ISBN 978-1-55365-531-2 (pbk.)
ISBN 978-1-926706-52-8 (ebook)

Cover and interior design by Jessica Sullivan
Cover bee illustration by Laura Kinder
Photos by Zachary Huang, except p. 108
Printed and bound in China by C&C Offset Printing Co., Ltd.
Text printed on acid-free paper
Distributed in the U.S. by Publishers Group West

We gratefully acknowledge the financial support of the Canada Council
for the Arts, the British Columbia Arts Council, the Province of British Columbia
through the Book Publishing Tax Credit, and the Government of Canada
through the Canada Book Fund for our publishing activities.

"End Notes for a Small History" by Betty Lies, originally published in
Southern Poetry Review 38 (Summer 1998), is reproduced by permission of the publisher.

"Angel of Bees" originally appeared in *Inventing the Hawk* by Lorna Crozier © 1992
and most recently appeared in The Blue Hour of the Day by Lorna Crozier © 1997.
Published by McClelland & Stewart Ltd. Used with permission of the author.

MIX
Paper
FSC FSC® C008047

To the Reader

The chiefest cause, to read good bookes,
That moves each studious minde
Is hope, some pleasure sweet therein,
Or profit good to finde.
Now what delight can greater be
Then secrets for to knowe,
Of Sacred Bees, the Muses Birds,
All which this booke doth shew.
And if commodity thou crave,
Learne here no little game
Of their most sweet and sov'raigne fruits,
With no great cost or paine.
If pleasure then, or profit may
To read induce thy minde,
In this small treatise choice of both,
Good Reader, thou shalt finde.

CHARLES BUTLER, THE FEMININE MONARCHIE,
OR A TREATISE CONCERNING BEES
AND THE DIVINE ORDERING OF THEM, 1609

CONTENTS

Introduction: Little Things *1*

{ 1 } BEES OF THE WORLD *7*

{ 2 } BEES AT HOME *33*

{ 3 } BEES OF THE FIELD *59*

{ 4 } LIFE LESSONS *83*

Acknowledgments 111
Notes 112
Selected Resources 116
Picture Credits 122
Index 123

LITTLE

THINGS

INSECTS:

"the little things that run the world."

E.O. WILSON, 1987

If you are like me, learning about bees will change your life. I'm not suggesting that you'll drop everything and devote yourself to studying insects (though that is possible). What I have in mind is more subtle: a new alertness, a quickening of wonder. Little things that, in the past, have slipped by almost without notice will now demand that you stop and pay attention to them. The hum of wings: whose wings? An insect darting among the flowers: is it a bee or a beefly, a bumblebee or a wasp? What is it doing? Where is it headed? True, it may take you a bit longer to water the petunias or pick the beans, but in those few stolen minutes, you will have been on safari. Gradually, you will begin to sense that a garden is not just a bunch of plants set out in pots and rows: it is a world within a world, a half-tamed ecosystem, full of some of the most exotic and astonishing creatures on the planet.

As children watching our first bees, we instantly catch the buzz. No one has to tell us that these bright careening atoms of life are miraculous.

> *➤ A goddess in the form of a bee was*
> *venerated at Camiros on the Greek island of Rhodes*
> *almost three thousand years ago.*

They are the hum of summer and the warmth of honeyed sun, but watch out: they also pack a punch. Right there, pressed headfirst into a blossom, beauty-with-six-legs meets danger meets the mystery of life. Bees are so much more than common or garden insects.

There is poetry to bees, and over the years, they have inspired many lovely thoughts. In ancient times, bees were said to be vestiges from a golden age or the only animals to have survived unchanged from the Garden of Eden. Some thinkers, no doubt licking bee-made honey from their fingertips as they spoke, imagined its sweetness raining down from the stars to the flowers, where the bees could gather it. ("Whether this is the perspiration of the sky or a sort of saliva of the stars," Pliny the Elder wrote in his *Natural History,* ca. 77 BC, "nevertheless it brings with it the great pleasure of its heavenly nature. It is always of the best quality when it is stored in the best flowers.") The poet Virgil wondered if some "portion of the divine mind" might be found in bees or if they were undying

BEE **WISDOM**

IT IS SAID by the Rabbis that when the Queen of Sheba visited King Solomon she held out to him two wreaths, one of artificial, the other of genuine flowers. Unable to discover which were the real flowers, the king caused the casement to be opened, when, lo! A flight of bees entered and lighted on the natural flowers.

So the bees were wiser than the wise king!

E.J. JENKINSON,

THE UNWRITTEN SAYINGS OF JESUS, 1925

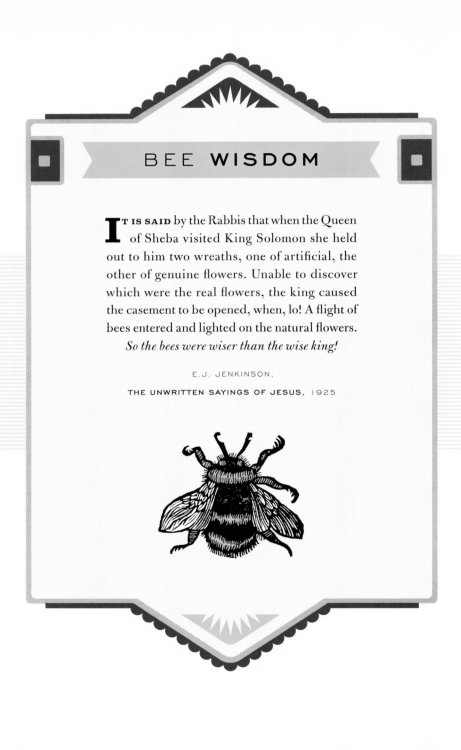

souls, soon to return to their celestial homes. And if bees were not themselves immortal, perhaps they were messengers sent by the gods to show us how we ought to live, in sweetness and in beauty and in peacefulness. All in all, it's not a bad rep for a bunch of stinging insects.

These days, I can hardly get out the door without being drawn off course by some small flying speck of life. Half the time, whatever-it-was has whizzed off before I get close enough even to hazard a guess at identification. Yet as much as I'd love to attach a name and a story to everything I see, I also enjoy the openness of not knowing. It reminds me that the childlike pleasure of amazement is always close at hand, as close as the nearest insect on a flowering plant. By bringing bees into focus, we open our eyes to the wonderful strangeness of the life that goes on, every day, all around us.

As we journey into the beeosphere, our main guides will be two of the most ingenious and audacious scientists you are ever likely to meet: the late Austrian zoologist Karl von Frisch and his protégé, German biologist Martin Lindauer, now in his 90s and living in Munich. As I became acquainted with the intertwining strands of their life stories, I was struck not only by their breathtaking discoveries—you'll see what I mean—but also by the human context in which they were working. Through the horrors of the Third Reich and World War II, these two men managed to honor beauty and peace by pursuing their studies of social insects. Later, when I spoke to American geneticist Gene E. Robinson and learned how the memory of the Holocaust had inflected his life work, I began to intuit that bees had brought light into a very human darkness. That light deserves to be honored, and so I have done my best to let it shine through in these pages.

1

BEES

of the

WORLD

The life of bees is like a magic well.

The more you draw from it,

the more there is to draw.

KARL VON FRISCH, 1886–1982

•   •   •

On a dark winter morning in 1945, a handsome twenty-seven-year-old biology grad student, and bee expert in the making, named Martin Lindauer found himself biking through the bombed-out ruins of a Munich suburb. As he pedaled through the rattling streets, he must have been reminded of other mornings in earlier, less troubled times, when he was a farm boy in the foothills of the Bavarian Alps. Perhaps he recalled the contentment he had felt then in the stony pastures and steep hayfields of his family's farm and, especially, among the bees in the apple orchard. Even as a small child, he had loved to sit beside the hives, watching the insects come and go: bees zooming in from distant meadows; bees flying loop-de-loops and spirals around the hive and darting back in through the door; bees carrying heavy loads of yellow or orange pollen. If he lay in the grass and gazed upwards, he could see a thousand bees at once, crossing and crisscrossing above him like shooting stars.

Bees brought sweetness out of chaos. Humans, on the other hand, seemed to have an instinct for devastation. Could it be that these insects have something to teach us?

Reality reasserted itself abruptly as Lindauer cycled along, when he misread the shadows and crashed his bike in a bomb crater. On foot and

Sic nos non nobis mellificamus apes. | Omnia in libris

All plants yeild honey as you see
To the Industrious Chymick Bee

then aboard a coal train, through air-raid sirens and delays, he continued toward his destination, the country retreat of the great Austrian zoologist Karl von Frisch, the world's leading expert on bee behavior and physiology. As director of the Zoological Institute in Munich, von Frisch had fallen under the suspicion of the Nazis as a person of "mixed descent in the second degree" (there were official queries about his maternal grandmother's genetics); yet despite persistent attempts to force him out of his position, he had somehow managed to maintain his classes, his research hives of bees, and his passion for discovery. To a young man like Lindauer—himself an anti-Nazi and an invalided conscript just home from the Russian front—von Frisch's biology classroom in Munich had been an oasis of light and hope in a world of madness.

⌃ Busy as bees, the scholars in this seventeenth-century woodcut add to the store of knowledge. The Latin motto reads, "So we the bees make honey but not for ourselves."

In the fall of 1944, when von Frisch evacuated his bee research program to the obscure village of Brunnwinkl in Austria, the last traces of decency seemed to flee from the city with him. And so it was, weeks later, that Martin Lindauer found himself traveling through the darkness, back into the brightness of bees. He had no way of knowing what a strange and beautiful world he was entering.

· · · The Flower's Little Friends · · ·

Anyone who takes an interest in bees, whether as a student like Lindauer or as a free and inquiring mind, needs to start with a review of the basics— a kind of Stinging Insects 101—before getting up close and personal with them. What is a bee, exactly? How many species are there? What is it about these insects that, for so very long, has made them fascinating to humans?

As Professor von Frisch once pointed out, "bees are as old as the hills." When the first *Homo sapiens* woke up and smelled the roses some thirty thousand years ago, bees had already been going about their business for well over 100 million years. ("This [antiquity] may be one of the reasons why they appear to be so mature," von Frisch noted, and "so perfect in many ways.") The oldest known fossilized bee is a tiny relic, about the size of a grain of rice, that was unearthed in Burma in 2006. Completely encased in amber, it dates from the early Cretaceous period, when dinosaurs like *Brachiosaurus* were still stomping through the swamps, and the gloomy coniferous forests were, for the first time, showing the colors of flowering plants.

TELLING THE BEES

*According to custom, honeybees will swarm
out of the hive and vanish if they are not told of a
death in the family and permitted to mourn.*

(For Edward Tennant)

Tell it to the bees, lest they
Umbrage take and fly away,
That the dearest boy is dead,
Who went singing, blithe and dear,
By the golden hives last year.
Curly-head, ah, curly-head!

Tell them that the summer's over,
Over mignonette and clover;
Oh, speak low and very low!
Say that he was blithe and bonny,
Good as gold and sweet as honey,
All too late the roses blow!

Say he will not come again,
Not in any sun or rain,
Heart's delight, ah, heart's delight!

Tell them that the boy they knew
Sleeps out under rain and dew
In the night, ah, in the night!

KATHARINE TYNAN, 1918

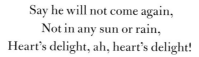

Bees evolved from wasps. To this day, some species of bees are so wasplike, and some wasps so like bees, that it is easy for a layperson to get confused. For those in the know, however, the two groups are distinct. Wasps—including the familiar paper wasps, yellow jackets, and hornets—are predators that kill other insects and spiders, by stinging, as food for their young; they also have a taste for carrion and, as many picnickers can attest, for burgers and fried chicken. Bees, by contrast, are

⌃ *Two forest beekeepers are caught in the act of stealing honey from wild honeybee nests, rousing the bees to a stinging defense.*

herbivores and use their stings only for defense. Instead of flesh, they raise their larvae on protein-rich pollen from plants. As far back as the Cretaceous, the tiny bees of Burma were already equipped with fringes of branched hairs, the bee tribe's signature adaptation for gathering pollen from flowers.

Bees and blossoms are made for each other, like the lovers in a romantic song. Biologists refer to this attunement as co-evolution, the long, slow, adaptive interplay among species that could not survive alone. For if bees depend on flowers for food, many flowering plants rely on animal helpers to reproduce, employing them as animated dildos, or sexual go-betweens, to transfer pollen from the ripe anthers of one flower to the receptive stigma of the next. Although butterflies, beetles, and other insects also provide this service, bees—with their hairy bodies and floral fixation—are especially suited to turn the trick and are the most important pollinating force on the planet. You could think of them as the flower's little friends, a dawn-to-dusk call-out service with a high approval rating from its clientele.

· · · *Bee Biodiversity* · · ·

For organisms with such a clearly defined mission—go forth and pollinate—bees are a surprisingly diverse group. There are currently more than nineteen thousand recognized species, and the experts confidently predict that as old taxa are reexamined and new specimens are brought in, the tally will soon exceed twenty thousand. (The classic who's who of modern bees, a tome entitled *The Bees of the World*, weighs twice as much as an average Bible.) The smallest living bees on record are no bigger than

A **STORE OF HONEY** was found in a hollow tree, and both the Bees and the Wasps declared that it was theirs. The argument grew so heated that the parties decided to put the case before Judge Hornet, the presiding authority in that part of the woods.

Witnesses told the court that they had seen striped, winged creatures near the hollow tree, humming loudly like Bees. Counsel for the Wasps insisted that this description fit his clients exactly.

Such evidence did not help. Finally, a wise old Bee made a suggestion. "I move," he said, "that the Bees and the Wasps be both instructed to build a honey comb. Then we shall soon see to whom the honey really belongs."

The Wasps protested loudly. Wise Judge Hornet quickly understood why they did so: They knew they were not up to the challenge.

"It is clear who made the comb," the Judge said. "The honey belongs to the Bees."

Moral: *Ability proves itself by deeds.*

fruit flies; the largest would fill half your palm. Although most are drab—the little black-and-brown-striped jobs of the insect world—some species are black as ants, and others are flying sparks of bronze, blue, or green iridescence. For reasons that aren't immediately obvious, the diversity of bees is greatest in arid regions (the Mediterranean basin is a hot spot, for example, as are the southwestern deserts of the United States), but most places are home to more kinds of bees than most of us would ever guess.

Since the days of the Cretaceous flower-wasps, the bee lineage has diversified into nine branches, or families. Each has a polysyllabic scientific name with no English equivalent and a description that draws attention to microscopic aspects of insect body structure. Just in case it should ever come up in conversation, the major extant families are the Colletidae (very roughly, plasterer bees, miner bees, masked bees), Andrenidae (small miner bees, dagger bees), Stenotritidae (colletid-like bees), Halictidae (sweat bees, alkali bees), Melittidae (including oil-collecting bees), Megachilidae (leaf-cutter bees, mason bees), and Apidae (bumblebees, carpenter bees, stingless bees, and honeybees).

The last two families in this list, Megachilidae and Apidae, are often grouped together as "long-tongued bees" not, as you might expect, because their tongues are especially long but because they have a longish fused sheath at the base of their mouth parts. Its function is unclear, though it may give these bees enough pucker power to suck up nectar from larkspurs and other deep-throated flowers. With or without this attribute, the various "long-tongued" and "short-tongued" species sport hairy, folding proboscises of various lengths, a feature that determines the types of flowers from which each species is able to drink.

With a couple of minor exceptions, most families of bees are represented on most of the flower-friendly continents. For obvious reasons, there are no bees in Antarctica. North America, by contrast, is home to six families and about 4,000 species—or about 4,001, if you count the honeybee, which was introduced from Europe and Africa. The state of

‹ *This illustration graced the cover of* The Humble-Bee, *1912.*

New York alone boasts 477 types of bees, and chances are that a roughly similar number are buzzing around in the countryside beyond your back door if you live anywhere in the north temperate zone. Precise figures are difficult to come by in many localities, especially in Europe, because of what one biologist has referred to, in upper-case distress, as a critical lack of study, or Taxonomic Deficit.

· · · *Life Stories* · · ·

When you get down to the nitty-gritty, insect identification requires technical knowledge and is a job best left to those who know their subantennal suture from their labial palpus. Fortunately for the rest of us, however, anyone with a basic grasp of reality can quickly acquire a sense of the many and varied faces of bee-ness. "This one is hairy and yellow; that one is shiny and green. How fascinating." To avoid embarrassing mix-ups, it is useful to keep in mind that bees have four wings, two fore and two aft, while flies, which may also feed in flowers, have only one set.

Apart from the flower garden, it is a general rule of thumb that bees are most likely to be discovered where we least expect to find them. Unlike the honeybees with which we are most familiar, the vast majority of bees are reclusive and solitary. Depending on the species and the circumstances, they make their nests in narrow cavities inside hollow stems, in rotten wood, or, most often, in the ground. Just look for the most dried out, neglected piece of soil you can find—the closely mowed edge of a roadway, for example, or the stony, sun-baked median in a parking lot—and there's a chance you'll find them there, roving over the ground like mini-helicopters. If you're out for a walk on the dark side, bedraggled

grass around headstones is another good spot to check. The less often a place is disturbed, the better the bees will like it.

Although the life history of these "digger bees" varies, they generally overwinter as either fully formed, mated females (the males having died off in the fall) or as late-stage larvae of both sexes that emerge and mate as spring adults. In either case, each female prepares (or at least renovates) a nest or nests for her young, often a straggling, descending tunnel in the earth, with tiny chambers, or nursery cells, budding off it. Alternatively, the chambers may be positioned end to end, with partitions of plant

⌃ *A dozen species of bees and wasps were pictured in* Blackie's Encyclopaedia, *around 1880. The bee at bottom right is digging a nest.*

resins, pebbles, leaf pulp, or mud in between them. Whatever the setup, the mother bee typically lines each cell with waterproof secretions, provisions it with a generous supply of pollen, lays a soft, whitish egg on top, seals the entrance, and leaves her offspring to eat its way to adulthood.

Sometimes hundreds of bees nest in the same patch of ground, riddling the soil with holes and forming a hazy, low-flying cloud as they zoom to and fro. But although these aggregations may look sociable, each bee is still very much on her own. Alone in the crowd, she devotes the few short weeks of her life to providing food and shelter exclusively for her own young, although she is unlikely to meet them. By the time the eggs hatch and complete their metamorphosis through four or five larval stages to become pupae and finally emerge as adults, the mother bee will have died or taken off. Very rarely, a small group of female digger bees share a common nest and, even more rarely, work together to prepare and provision cells, but most of the time, most bees are loners.

· · · The Sisterhood · · ·

Most but, of course, not all. A relatively small fraction of bees, around 4 percent of the total, are social. This number includes members of several common groups, including many sweat bees (smallish, ground-nesting bees, often brightly iridescent, with a namesake taste for salt) and carpenter bees (largish, hairy insects that excavate nest galleries in

‹ *A sparky little sweat bee, of the family*
Halictidae, probes the sweetness of an aster.

wood), as well as the majority of bumblebees (the fuzzy teddy bears of the bee world).

Again, details vary, but in the case of the bumbling humble-bee, the story begins in early spring when a mated female, the only member of her tribe to survive the winter, comes out of hibernation and searches for a safe place to rear her brood, often in a tussock of dead grass or an abandoned mouse burrow. In this snug den, she builds a little wax cup, or honeypot, and fills it with floral nectar, as a larder to help sustain her during her labors. And such labors: not only does she gather and shape a large ball of pollen, lay her eggs, four to sixteen in a batch, and cover the whole thing with wax, she actually lies atop the ball and incubates. To do so, she shivers her large flight muscles (without vibrating her wings), thereby generating enough heat to keep the eggs at a cozy 85°F.

This maternal care is rewarded four or five weeks later, when her offspring—all females—emerge as adults and set to work foraging and helping around the nest. Thereafter, the fertilized mother bee, or queen, devotes herself primarily to egg laying, while her unfertilized daughters, the workers, help to rear successive broods of their sisters. At its height, a successful bumblebee colony will number between thirty and four hundred female members. The males, or drones, have only one interest in life and are produced in late summer on a just-in-time schedule to mate with fresh young queens for the following year.

It is oddly pleasurable to realize that any time we see a bumblebee nosing around the garden, there are dozens of others just like her in some old mulch pile or under a neglected hedge, sharing in the daily cares of

THE **HUMBLE**-BEE

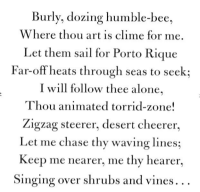

Burly, dozing humble-bee,
Where thou art is clime for me.
Let them sail for Porto Rique
Far-off heats through seas to seek;
I will follow thee alone,
Thou animated torrid-zone!
Zigzag steerer, desert cheerer,
Let me chase thy waving lines;
Keep me nearer, me thy hearer,
Singing over shrubs and vines...

Wiser far than human seer,
Yellow-breeched philosopher!
Seeing only what is fair,
Sipping only what is sweet,
Thou dost mock at fate and care,
Leave the chaff, and take the wheat.
When the fierce northwestern blast
Cools sea and land so far and fast,
Thou already slumberest deep;
Woe and want thou canst outsleep;
Want and woe, which torture us,
Thy sleep makes ridiculous.

RALPH WALDO EMERSON, 1839

communal life. Bee societies are families. And this is true not only for species like bumblebees, with their seasonal colonies, but also for two other
extravagantly—even bizarrely—social tribes of bees. They are, first, the
stingless bees, a group of about 120 species of mostly small, black, flower-
feeding insects that are found in the tropics of the Americas, Africa, Australia, and Asia. Despite their reassuring name, stingless bees are not
biteless, and one observer reports that they "swarm over the heads of
humans who venture too close, locking their jaws so tightly onto tufts of
hair that their bodies pull loose from their heads when they are combed
out." Some species also spew caustic secretions that blister an intruder's
skin, a trait that in Brazil has earned them the name "fire shitters."

From a careful distance, however—say, halfway around the globe— the most remarkable trait of stingless bees is not their capacity for collective defense but their incapacity, as individuals, to take care of themselves. Stingless bees live in family colonies of up to 100,000 individuals, most of which are again female. But unlike bumblebee females, which are all essentially the same, stingless queens and workers are anatomically specialized. Queens are unable to forage; workers are unable to mate. If most of the world's bees are "all about me," individual stingless bees are part of a larger whole and their devotion to the sisterhood is not negotiable. Biologists refer to them as "highly eusocial" insects, a term that technically refers to societies in which members of two generations (a fertile queen and her physiologically sterile daughters) cooperate to rear the mother's young. The term translates literally as "highly" or "beautifully social," which just about says it all.

· · · *Sweetness and Light* · · ·

The remaining group of beautifully social bees is the honeybees, genus *Apis*. Here, the beauty lies not only in the bees' social order but also in their nests, those exquisite hexagonal constructions of thinnest beeswax. Although honeybee taxonomy remains a work in progress, the experts tell us that there are at least nine species, mostly in India and Asia, and

‹ *In this array of British bumblebees, the queens of six species appear in the top row, with workers in the middle and drones at the bottom.*

that more may be hiding in the jungles of Borneo and on islands in the South Pacific. As a group, honeybees are sleek little things, with slender bodies, elegant black or golden stripes, and an overall preference for tropical climes. Some, including several species of "dwarf" and "giant" honeybees found in Southeast Asia, nest on single, open combs attached high up in trees, protected by a shimmering layer of ever-moving bees. The others are cavity nesters that build elaborate nests in caves and hollow trees or in the confines of a cozy apiary.

The best-known member of the genus—indeed the best-known insect in the world—is *Apis mellifera,* a species so prized by humans that it has eclipsed all the rest and usurped the name of "bee" exclusively for itself. Originally native to Africa, the Middle East, and western Europe (from the Cape of Good Hope north, through savannahs, rain forests, and gardens, to southern Sweden and from the western Sahara east into Iran), the species is now found at temperate and tropical latitudes virtually around the world, thanks to the exertions of its human admirers. In 1622, for example, English colonists bound for Virginia bundled up their woven-straw skeps, or hives, and transported them by sailing ship across the Atlantic; these were the first honeybees to make the voyage. By 1800, feral honeybees, or "Englishman's flies," as Native people

➤ *Under the curious eye of his children, a nineteenth-century beekeeper transfers his bees from old-fashioned straw skeps (the beehive-shaped objects in the shed) to a more modern box hive.*

sometimes called them, had crossed the Mississippi, often in advance of settlement, and rapidly made themselves at home across the continent. A similar story unfolded in Australia in the mid-nineteenth century, after honeybees were introduced there in the 1820s.

Honeybees are sweetness and light—producers of honey and wax—so it is no great wonder that the colonists packed them with their baggage. (No other species can equal *Apis mellifera* for its honey production.) And the insects offer another gift that would have been hard to give up: the quiet contentment of the beeyard. "And I shall have some peace there," Yeats wrote, "for peace comes dropping slow." This is the peace of a thousand thrumming wings, the peace of a great community engaged by a common goal, the peace that Martin Lindauer had sensed as a child. Like stingless bees, honeybees live in teeming cities of tens of thousands of mostly female residents (a queen and her daughter-workers), who somehow find a time for every task, and every task in its season. This harmonious interaction hangs over the beeyard like the scent of honey, a sweet invitation to study.

· · · *A Bee in His Bonnet* · · ·

Karl von Frisch was an austere-looking man, with a large nose, high, bony forehead, and the wire-rimmed glare of an exacting taskmaster. But how he did love his bees. When someone asked him in later years how he felt about receiving a Nobel Prize for his honeybee research (an award he shared with animal behaviorists Niko Tinbergen and Konrad Lorenz in 1973), he acknowledged that the honor had "made me very happy."

"But I have to say with all honesty," he reflected, that "when I am in the yard outside observing. . ., and I hear the buzzing of the bees, that is for me a greater experience than the Nobel Prize."

When von Frisch began his honeybee studies in the early 1900s, many large questions were beginning to buzz drowsily through the minds of zoologists. For instance: Given that the great majority of bee species are solitary, why had a few, like the honeybee, become so intensely social? And why was it that these social bees could live together in harmony and peace, when humans often failed to do so spectacularly? But these were subjects for meditation, not topics for research, and von Frisch focused his energies on more practical matters. When his research into small questions began to turn up large answers, he was as surprised as anyone.

It had all begun at a meeting of the German Zoological Society in Freiburg in 1914, when a bumptious young von Frisch presented a paper entitled a "demonstration providing experimental evidence for the exis-

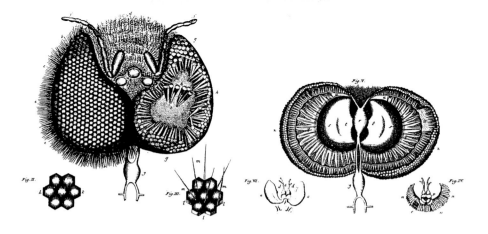

tence of color vision in animals supposed to be color-blind." With this title, he was thumbing his nose at a respected scientist who had recently declared that insects and other "lower" animals were unable to perceive color. But when von Frisch looked out the window and saw gardens of brightly painted flowers with bees dancing around them, he knew this supposition had to be false. Bees must perceive colors, but how to prove it?

Von Frisch's solution to this problem was to organize a kind of honey-bee circus. First he trained his performers by setting out dishes of sugar water on squares of colored paper—usually an alluring shade of blue—and allowing some of his honeybees to feed from them. Thereafter, if he set out a scrap of blue paper, the trained bees would flock to it even if there was not the slightest trace of sweetness anywhere around it. For his demonstration to the Zoological Society, von Frisch upped the ante by displaying an array of squares in which the blue patch was surrounded

↖ *These magnified views of the eye of a honeybee date from the 1600s. Left, the three round ocelli, flanked by the huge compound eyes. Right, a cross-section showing the optic nerve.*

THE **ORIGINS** OF HONEYBEES

The god Re wept and the tears
From his eyes fell on the ground
And turned into a honeybee
The bee made [his honeycomb]
And busied himself
With the flowers of every plant;
And so wax was made
And also honey
Out of the tears of the god Re.

A RELIGIOUS TEXT FROM ANCIENT EGYPT

on all sides by a checkerboard of grays. If honeybees really were color-blind, the blue would appear gray to them and blend in with the rest, and the demonstration would end in confusion. But von Frisch's troupe of bees performed "as if by command," crowding onto the blue squares, avoiding all the rest, and even seeking out a blue necktie in the audience.

Honeybees could see color. But this result, as satisfying as it was, left von Frisch with more questions than when he had begun. Which colors could bees see? Did they also learn and remember scents? And there was one chance observation that had von Frisch scratching his head. In performing his experiments, he had sometimes left an empty food dish on the table. Although most of his bees soon lost interest and stopped coming around, occasionally one or another of them would drift by and land. If the dish had been refilled and the scout returned to the hive with a full crop, "the whole company of foraging bees was buzzing round the dish again within a few minutes."

"It was clear to me that the bee community possessed an excellent intelligence service," von Frisch observed, "but how it functioned I did not know." What had begun as a straightforward investigation—could honeybees see color? yes or no—had suddenly opened out into a world of speculation. From then on, the mysteries of honeybee society would, as von Frisch put it, give him no rest. Thirty years and two world wars later, when an aspiring bee researcher named Martin Lindauer appeared at his door, Professor von Frisch was more than happy to put a bee in his bonnet.

2

BEES

at

HOME

... this being an Amazonian

or feminine kingdome

CHARLES BUTLER, **THE FEMININE**

MONARCHIE, OR A TREATISE CONCERNING BEES

AND THE DIVINE ORDERING OF THEM, 1609

•   •   •

When Martin Lindauer returned home to Munich from his consultation with Professor von Frisch, he had an assigned topic for his honeybee research but no place to research it. The Zoological Institute, with its beeyard—once von Frisch's pride and joy—had been reduced to rubble, collateral damage in the war against the Third Reich. For a time, Lindauer and some of his fellow students had lived in the upper stories of the buildings, where they could fight fires lit by the bombs, but as the Allied bombardment intensified, they had been forced to abandon their posts, retreat to lower floors, and eventually hole up in the cellars. In the end, an explosive intended for the nearby train station struck the institute instead and left the structure so unsound that even this refuge had to be abandoned.

Fortunately for Lindauer and for the future of bee research, he had recently married and, through his new wife, had acquired a house in the suburbs. Although it too had suffered damage, at least it had a roof and several habitable rooms in which the young couple could live. Best of all, it had a garden where he and his bees could work. For years afterwards, Lindauer would remember the surreal moment, on a sunny May afternoon in 1945, when he looked up from tending bees in his yard to see a

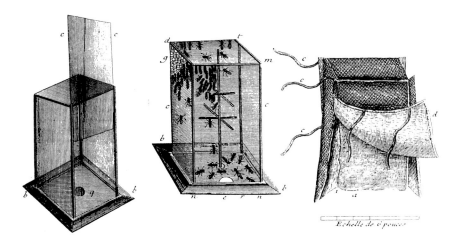

Echelle de 6 pouces

cavalcade of American tanks rolling past the house. The rumble of war was fading into the distance, and the humming of bees filled the garden.

And so Lindauer set to work. Although the topic that von Frisch had assigned him was not particularly interesting in itself, just sitting and watching the insects was pure satisfaction. ("After experiencing the senseless regime of the Hitler time," he commented later, "I drew strength from having work based on absolute correctness, honesty, and objectivity... I found a new home with the bees. It really was a new home, the bee colony.") Whatever the specific results of his study, it was certain to be a step toward the larger goal of understanding how a honeybee colony works. To this end, Lindauer spent countless hours in his garden, getting better acquainted with his research subjects.

⌃ *In the 1700s, scientific observers began to house*
their bees in glass boxes like these, for easier observation.

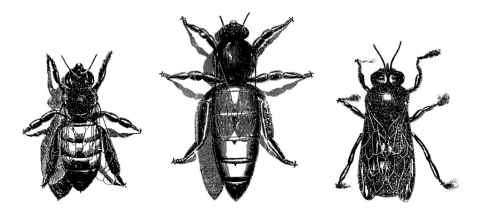

Rather than enclose his colonies in conventional boxes, with opaque walls and tops, Lindauer followed von Frisch's example and housed them between panes of glass. Each frame in the hive was constructed like a sandwich, with glass for bread and a double-sided sheet of honeycomb as filling in the middle. There was just enough space under the glass for a bee to walk across the comb, so that all the inhabitants were always exposed and visible. Then, rather than slotting the combs one behind another, as an ordinary bee-keeper would, the young scientist arranged them side by side and end to end, like the glass in a multipane window. If such an observation hive was provided with shelter, the bees seemed happy with it, and a person could sit and watch them to his heart's content.

· · · The Queen Considers · · ·

What did Lindauer see when he peered through this magic window into the bees' private world? A whole lot of insects running around in every direction. "Anyone who has ever looked into a beehive," he once noted, "has surely asked himself, can there be any positive collaboration in that confusing, teeming mass of thousands of bees? But this collaboration must exist; otherwise the building of combs, the raising of brood, the collection and storing of food could never be carried out so regularly and harmoniously."

The heart of the whole operation is, of course, the queen, and if you search the frames of the observation hive patiently, you'll eventually locate her, walking slowly, even majestically, across the honeycomb. She looks more or less like a regular bee—a little bigger, with a longer abdomen and, sometimes, a paler, honey-gold hue—but what really sets her apart from her hive-mates is the way that they treat her. Everywhere she goes, she is accompanied by a retinue of attendants (attentive daughters) that nudge her and nuzzle her, feed her and groom her, and stroke her with their antennae. And always, she is on the same monotonous maternal mission. On any given summer day, a queen bee typically lays about 125 dozen eggs, with a combined mass equal to her own. That's 1,500 eggs per day or 225,000 per breeding season or about half a million in the two years of her expected life span.

Each of these thousands of potential progeny receives careful consideration. First, the queen chooses a cell in the honeycomb (typically in the central portion of the frame) and sticks her head and forelegs in, as if making an inspection. Then, if she finds the cell to her liking, she wheels around, curves her long abdomen into the cavity, deposits an

◁ *From left to right, the honeybee worker, queen, and drone,*
as pictured in A.I. Root's classic, The ABC and XYZ of Bee Culture, *1908.*

BEHOLD THE **KING**

For centuries, many learned authorities
preferred to believe that the queen bee was actually a king
with an inexplicable predilection for egg laying.
This account comes from Book 11 of Natural History
by Pliny the Elder, published around 77 BC.

THE KINGS have always a peculiar form of their own and are double the size of any of the rest; their wings are shorter than those of the others, their legs are straight, their walk more upright, and they have a white spot on the forehead, which bears some resemblance to a diadem: they differ, too, very much from the rest of the community, in their bright and shining appearance...

. The obedience which his subjects manifest in his presence is quite surprising. When he goes forth, the whole swarm attends him, throngs about him, surrounds him, protects him, and will not allow him to be seen. At other times, when the swarm is at work within, the king is seen to visit the works, and appears to be giving his encouragement, being himself the only one that is exempt from

work: around him are certain other bees which act as body-guards and lictors, the careful guardians of his authority...

When they are on the wing, every one is anxious to be near him, and takes a pleasure in being seen in the performance of its duty. When he is weary, they support him on their shoulders; and when he is quite tired, they carry him outright.

oblong egg, extricates herself, and moves on to the next and the next. What we cannot see is that these pearly white eggs differ in subtle ways. Some are fertilized, or diploid, and carry a double set of chromosomes, while others are unfertilized, or haploid, with only a single set. Thanks to one or more youthful mating excursions, the queen carries a lifetime supply of sperm (gleaned from as many as twenty males) inside a spherical organ called a spermatheca. The queen has the option, with each egg she lays, of dispensing sperm from this store or holding it back. This simple yes-no decision determines the sex of her offspring: fertilized for female, unfertilized for male.

· · · Drones: Big, Hairy Guys · · ·

Whether the queen dispenses sperm or withholds it depends on the dimensions of the cell in which she has chosen to lay. Honeycomb consists of a lacy matrix of hexagonal chambers, each tooled by worker bees to precise specifications. Most measure 1/5 inch across, give or take a hair's breadth, but some—the deluxe rooms in the Honeycomb Hotel—are a good 20 percent larger. It is into these oversized chambers that the queen deposits her sons. (Apparently, she makes the size discrimination with her forelegs, when she inspects the cell. Remove these crucial appendages, as some researchers have done, and she loses her ability to make appropriate judgments.)

Thus royally lodged, the males have the extra space they need to make the transition into plump larvae, plumper pupae, and big, fat, hairy adults. And what princes they become. Unable to forage for themselves, adult drones constantly beg their worker-sisters for feedings or help

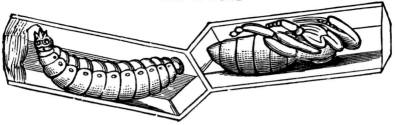

themselves to free rations from the colony's honey reserves. They may leave excrement around the hive, where their sisters clean it up, and take no interest in the family business. Even von Frisch judged drones to be "a little stupid and lazy." But when it comes to sex, these guys are on the job.

As soon as a drone is sexually mature, at about twelve days of age, he chooses a sunny afternoon in spring, grooms his antennae, cleans his huge eyes ("the better to see you with, my dear"), and flies out looking for action. Somehow or other, he manages to hook up with other aspiring males in a traditional congregation area, or aerial pickup place, and they all hang out together, waiting for their chance. When a young unmated queen shows up on a nuptial flight, the heretofore slothful drones, now drunk on her pheromonal perfume, take off after her as fast as their large wings and powerful flight muscles will take them. Mating occurs on the wing at speeds of up to twenty miles per hour. If a male doesn't make contact, he'll return to his hive (or to the wrong hive, ninny that he is)

↖ Sealed inside its brood cell by a cap of wax, the honeybee larva on the left has spun a cocoon, from which it will emerge eight or nine days later as a perfect adult, like the one on the right.

and come out on another day to try his luck again. But if he succeeds in mating, his glory is also his doom. The queen tears away from him with his penis and viscera dangling from her rear, and the spent drone tumbles, dying, out of the air.

The outcome is not much better for the drones that find their way home. By high summer, the nectar flow to the colony is declining, all the young queens have flown, and the males remaining in the hive find themselves *apis non gratae*. The workers, which up until now have tended their brothers with care, suddenly begin to harass them, by biting their bodies and plucking at their wings. Sometimes, a worker will grab hold of a drone that is twice her size and drag him to the door. As von Frisch says, "They could not make their meaning clearer." When things get especially nasty, the workers sting the drones to death, an assault that the stingless males are helpless to resist. Although a worker bee dies after attacking a mammal, because her barbed stinger gets caught, she can puncture the soft flesh of her brothers over and over. It's not all sweetness and light in Beesville.

· · · *Workers: Enlightened Self-Interest* · · ·

Worker bees are born to serve the greater good, whatever that good may be. If the colony needs drones, they'll tend them. If it needs to get rid of them, they'll kill. How they know what needs doing when is a deep mystery and one that Lindauer, watching the bees in his garden, loved to ponder.

‹ *This cheerful tribute to beelike virtue was created in 1929.*

There is nothing particularly remarkable about the life history of worker bees. By far the most populous caste in the colony, they begin life as fertilized, or female, eggs laid in standard-sized cells and, as larvae, are fed by their adult sisters on standard brood-feeding secretions. As a result of this modest upbringing, they develop into nonbreeding adults that are nonetheless fully equipped to care for offspring. Meanwhile, on the margins of the comb, a small number of other females are being reared in special capsules and fed a lavish diet of nutrient-rich secretions, or "royal jelly," a regime that triggers their transformation into queenly egg-layers. Together queen and workers are able both to produce and rear young, a combination that permits them to function together as one complete parent.

Through the glass walls of his observation hive, Lindauer could watch the workers as they scurried around performing their household tasks. Here, bees were cleaning out vacant cells in preparation for reuse, by removing old cocoons and recoating the walls with wax. Over there, others were poking their heads into occupied cells, the ones with grubs in them, to check on the larvae and see if they needed to be fed. (According to Lindauer's data, nurse bees inspect each larva, on average, 1,926 times during the five or six days before it makes its cocoon but feed it on only 143 of those visits.) Elsewhere in the hive, bees were busy building comb, capping comb, packing comb with pollen. Tucked away in a quiet corner, an individual might be flicking a droplet of nectar in and out on

> *Nurse bees tend a queen larva in her royal cell, or queen cup.*

BEES

For so work the honey-bees
Creatures that by a rule in nature teach
The act of order to a peopled kingdom:
They have a king, and officers of sorts;
Where some, like magistrates, correct at home,
Others, like merchants, venture trade abroad,
Others, like soldiers, armed in their stings,
Make boot upon the summer's velvet buds;
Which pillage they with merry march bring home
To the tent-royal of their emperor,
Who, busied in his majesty, surveys
The singing masons building roofs of gold,
The civil citizens kneading up the honey,
The poor mechanic porters crowding in
Their heavy burdens at his narrow gate,
The sad-ey'd justice, with his surly hum,
Delivering o'er to executors pale
The lazy yawning drone.

WILLIAM SHAKESPEARE, 1599

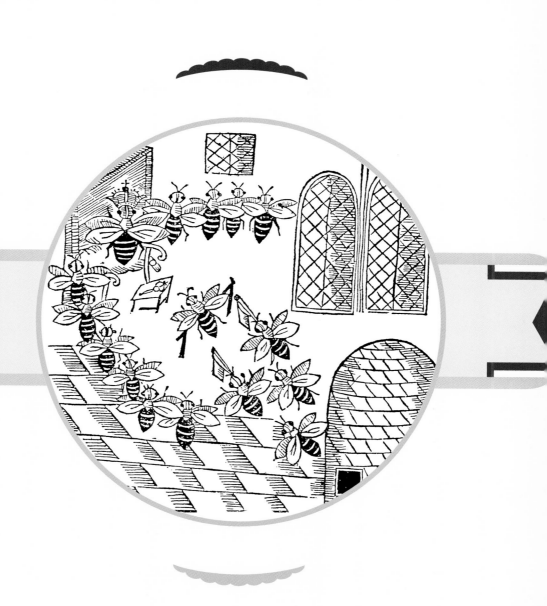

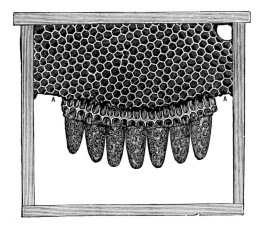

her tongue, waiting for the honeyed glob to thicken. At the same time, others were fanning their wings near the entrance, for cooling or ventilation, or standing guard in the doorway, with their forelegs raised and their antennae up, at attention.

It's a busy life, and worker bees can't take it for very long. Except for individuals that emerge in late summer and pass the winter inside the hive, workers seldom live for more than sixty days. By the time they keel over, they are typically faded and worn from exertion, having given their all in the service of their mother, their brothers, and their thousands of sisters. Biologists describe this situation as an extreme case of "kin selection," in which individuals pass on their genes indirectly, through their close relatives, rather than directly, through personal descendants.

Although this setup is odd by human standards—a squadron of spinsters that devote themselves selflessly to their sibs—the rules of bee society actually make perfect sense. Because of the asymmetries of the bees' gender-determination system, in which males are haploid and females are diploid, a worker shares three-quarters of her genes with each of her sisters (more than we share with our young), and so her apparent self-sacrifice is actually a kind of enlightened self-interest. She has much less in common with her brothers (only a one-quarter relationship), and per-haps this lack of consanguinity helps to account for her seasonal willing-ness to knock them off. She has even less time for her even more distantly related nephews—unfertilized eggs that, on the rarest of rare occa-sions, are produced by other workers—and these she nips in the bud by sniffing out the offending morsels and eating them. There are limits to sisterliness.

· · · *From Queen to Genes* · · ·

There is something uncanny about honeybee behavior, an aura of know-ingness that we may not expect to find in "mere" insects. How does a bee, with a brain the size of a flax seed, perceive a need to clean out cells where the queen will lay her eggs? How does a worker recognize which larvae need to be fed or when the honey is thick enough for storage or that rearing nephews is not in the community's best interests? And beyond that, how can a colony of sixty thousand or more individuals coordinate its activities amid all the bustle and scurry of an insect city? For centuries, from Aristotle right up to the threshold of modern times, most observers had assumed that the hive was governed by a benevolent dictator—an all-knowing king or queen—that understood the big picture and told the workers what to do. But one has only to watch the Royal

‹ *When they sense the time is right, workers prepare special vertically hanging cells for rearing queens, usually along the edges and bottoms of the combs.*

Ovipositer at her ceaseless task to know that the queen has no mandate for governance.

Mind you, a queen bee does exercise a certain subtle dominion that the ancients knew nothing about. Scientists tell us that she produces a kind of glandular elixir, known as the "queen substance," which her courtiers pick up on their antennae when they touch her and which they then distribute, on rapid zigzag runs, to the farthest corners of the honeycomb. When this pheromone is present, the workers go placidly about their tasks and seldom rear new queens. When it is absent (because the queen has died or been removed), the controls are off and the whole hive literally roars with discontentment.

Yet despite this ineffable presence, the queen is certainly not the boss. So how is the work of the hive governed? By the time Lindauer started his research, a profound new answer to this question had begun to emerge. Like all organisms on Earth, honeybees are living, breathing expressions of instructions encoded in their genes, instructions that have been refined, over an endless span of years, by the implacable logic of natural selection. (Echoes of von Frisch: "This may be one of the reasons why [honeybees] appear to be so mature, and so perfect in many ways.") If a new piece of genetic code improves the reproductive success of individuals that carry it, it will flow on into the future. If not, it will die out. And this unsparing rule applies whether the gene in question

> *A queen bee reclines in regal splendor in this playful drawing by*
Wilhelm Busch from his book Schnurrdiburr oder die Bienen Braun, *1869.*

controls a physical characteristic—human eye color, for example, or the structure of a bee's pollen-collecting hairs—or a finely tuned and adaptive behavior.

Could it be that honeybees are highly evolved robots, their every action and reaction dictated by the coiled complexity of their chromosomes? By the 1940s, this conclusion seemed inescapable not only in light of evolutionary theory but also in view of observations that had been made, some years earlier, by a man named G. A. Rösch, one of von Frisch's early students. Working with an observation hive, Rösch had gently captured worker bees as they emerged from their cells and, paintbrush in hand, had marked each of them for identification. (A white spot at the head end of the thorax stood for 1, a red spot for 2, and so on,

AGAINST **IDLENESS** AND MISCHIEF

This hymn for children is now most fondly remembered as the inspiration for Lewis Carroll's mischievous parody, "How Doth the Little Crocodile" in Alice in Wonderland.

How doth the little busy Bee
Improve each shining hour
And gather Honey all the day
From every opening Flower!

How skilfully she builds her Cell!
How neat she spreads the Wax!
And labours hard to store it well
With the sweet Food she makes.

In Works of Labour or of Skill
I would be busy too:
For *Satan* finds some Mischief still
For idle Hands to do.

In Books, or Work, or healthful Play
Let my first Years be past,
That I may give for every Day
Some good Account at last.

ISAAC WATTS, 1715

through five colors and three positions, a system that permitted him to number up to 599 individuals.) He then checked on his marked bees day after day and took note of what each was doing. By pooling all these data, he was ultimately able to show that honeybee workers typically perform an age-based sequence of tasks, from cleaning cells (roughly week 1, on average), to tending the queen and brood (week 2), to handling nectar and building comb (week 3), to guarding the entrance and foraging outside the nest (week 3 or 4 to death). As each stage is reached, glands for brood feeding or wax production develop on schedule, the appropriate behaviors kick in, and the whole colony runs with an efficiency that makes it the envy of mere humans.

· · · *Bee 107* · · ·

As Lindauer watched his own worker bees play out their lives behind glass, he had no doubt that the basic conclusions of this study were correct. On average, workers perform different tasks at different life stages, and as a result, the work of the hive is shared among community members. But he was irritated by Rösch's reliance on averages and norms and by the implication that bees progress through life on a rigid program. Real bees, the ones that Lindauer had devoted his life to observing, lived in a world of wind and weather where conditions were always in flux and where the needs of the colony changed along with them. Although he couldn't prove it, Lindauer had a hunch that something subtle must happen inside the hive, some sort of communication perhaps, that permitted workers to sense and respond to these shifting requirements.

And so began his acquaintance with Bee 107. She was an ordinary honeybee worker that Lindauer caught and marked on July 5, 1949, just as she was emerging into adulthood. From then until her death in a thunderstorm twenty-five days later, he sat and watched her for hours (sometimes into the night), patiently timing her activities with a stopwatch. On the whole, as he later reported, she followed the developmental sequence that Rösch had described, beginning by cleaning cells in the area of the hive in which she herself had emerged and later focusing her energies on feeding larvae. From the outset, however, her behavior also proved to be remarkably flexible. On the eighth day of her life, for example—when she "should" have been exclusively engaged in nursing young—she also found time to clean cells, cap pupae, build comb, and generally pitch in wherever she was needed.

What really surprised and fascinated Lindauer was the amount of time that Bee 107 spent walking through the hive, poking her head into corners, or visiting areas where comb was being built, for all the world as if she were looking for things to do. In his 177 hours of observation, she spent fully 56 hours just looking around, and in every case, she ended up finding something that needed to be done. Commenting again on Day 8, Lindauer wrote, "One can see that every tour of inspection results in some form of activity: nursing once, building or cleaning on other occasions, just as the need arises." The subtlety that Lindauer had sensed in

➤ *This diligent worker bee is part of a crew preparing honey for storage.*

the hive turned out to reside in Bee 107 herself and in each of her thousands of companions.

So how does a worker bee know what to do when? For an individual engaged with tasks inside the hive, as Bee 107 was, the answer turned out to be deceptively simple. She sees a job that needs to be done and that she is able to perform, and then she gets busy and does it. Although it is impossible to know what goes through her mind as she does this, or if bees have minds at all, it seems likely that most of her actions are

ᴧ *A forager skips into the hive with*
a load of nectar in this cartoon by Wilhelm Busch.

instinctive. When she encounters an appropriate stimulus—a certain sight or texture or smell—she makes an appropriate response, drawing on her large, inherited repertoire of genetically encoded behavior. "Thus we have found in division of labor [within the hive] a simple method of spreading information," Lindauer explained in his German-flavored prose. "It is not a mutual communication from bee to bee; instead, each bee is its own informant."

Finally, Lindauer had a partial answer to his questions about how the harmony of the hive is achieved: alert individuals act on their own to serve the community. But as satisfying as this conclusion was, it couldn't explain everything he had observed. There was the time during his surveillance of Bee 107, for example, when he had placed a heat lamp near the hive in which she and her colony lived, thinking to keep them warm, but had mistakenly set it too close for comfort. By the time he came back to resume his observations, overheated bees could be seen depositing water at the entrances of brood cells or flicking droplets in and out on their proboscises, cooling the hive by evaporation. Even supposing that the hive bees had responded to the crisis by instinct (an impressive accomplishment in itself), this didn't explain how they had suddenly obtained the water needed for cooling. Somehow, news of the emergency must have spread to workers outside the hive—the only ones that could bring supplies in—alerting them to stop foraging for pollen and nectar and to start delivering coolant. Was it possible that bees could not only recruit one another as foragers, as von Frisch suspected, but that they could also tell each other to switch tasks? Was there no limit to their sophistication?

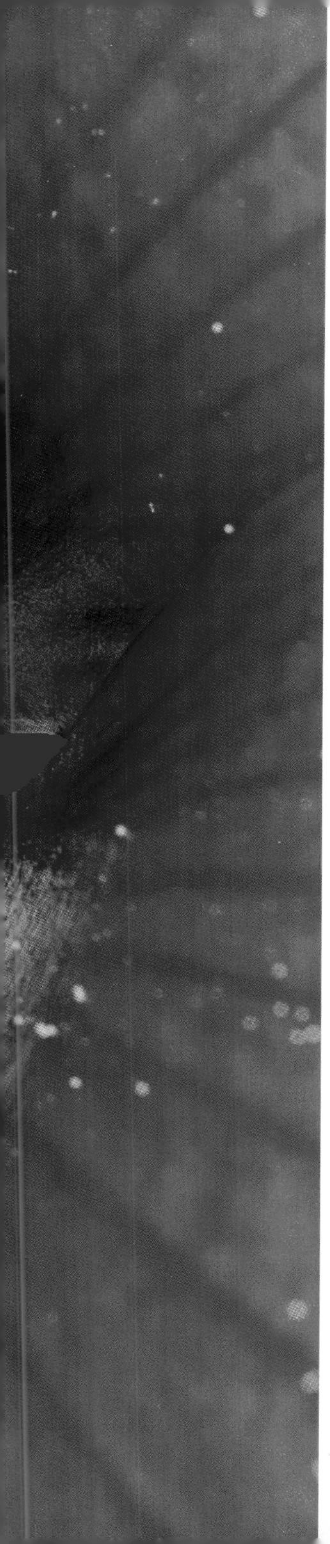

3

BEES

of the

FIELD

The fragrant work

with diligence proceeds.

VIRGIL, **AENEID**, CA. 29 BC

• • •

By the time Lindauer concluded his study of Bee 107, he had graduated from his backyard laboratory in Munich and accepted his first professional posting, as research assistant to Karl von Frisch in Austria. There, he found the great man in a fever of excitement. Over the preceding couple of summers, von Frisch had made observations so startling that he scarcely knew what to think. "I did not want to believe what I could see with my own eyes," von Frisch admitted; but facts were facts. The same bees that began their adult lives as inside workers, each more or less under her own command, grew up to become outside workers, or foragers, with astonishing gifts of communication.

When von Frisch had begun his studies thirty years before, both the questions and the probable answers had seemed simple. All he had wanted to know back then was how a lone scout that has found a rich source of nectar lets her hive-mates know about it, so that one bee becomes dozens or hundreds in a matter of minutes. By watching foragers as they returned to the hive, he had noticed that most re-enter quietly, look around for inside workers to suck up their nectar load, and then head straight out to gather more. By contrast, a forager that has

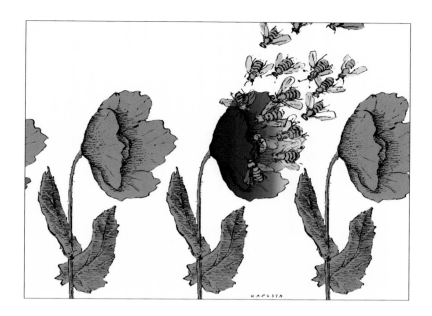

found something special—flowers brimming with super-sweet nectar or a dollop of honey on a scientist's glass dish—reacts very differently. She rushes into the hive and starts dancing.

· · · Slamdancing Bees · · ·

We're not talking ballet here; this is slamdancing, as the bee shoves into a cluster of unemployed foragers that are hanging around near the door and starts jostling among them in narrow circles. All six feet a-running,

^ *The drama of pollination becomes visible*
in this illustration by Polish artist Janusz Kapusta.

she twirls to the right and whirls to the left, like someone performing an impassioned dance. In no time, the excitement has spread through the mosh pit, as bees reach out to touch her with their antennae and trail after her, trying to keep in contact with her abdomen. In von Frisch's words, the followers "take part in each of her manoeuvrings, so that the dancer herself, in her madly wheeling movements, appears to carry behind her a perpetual comet's tail of bees." And so the performance continues for a few seconds or a minute more until the dancer breaks off, heads for the door, and flies back to her feeding place for another load. Soon, some of the dance-followers rush out of the hive, locate the bonanza as well, and return home to become dancers themselves.

The more dancing there is for a food source, the more bees are recruited to it. Somehow, in all the jittering and jiving, useful information is transferred, and for the longest time, von Frisch thought he understood how it worked. When a dancer bursts into the hive, he suggested, her flamboyant movements catch the notice of other bees, which crowd around her gyrating body. Once in contact, the followers rely on chemical receptors on their outstretched antennae to pick up the scent of flowers that adhere to the dancers' abdomen. With this information stored in memory, the new recruits leave the hive and fly in ever-widening circuits until, through luck or diligence, they discover flowers with matching attributes. In one of von Frisch's early experiments, bees

‹ *As they nuzzle into blossoms, bees pick up traces of floral scents. The bee on the right is carrying a packet of pollen on her corbicula, or pollen basket.*

LIKE **AND** LIKE

A fair bell-flower
Sprang up from the ground;
And early its fragrance
It shed all around;
A bee came thither
And sipp'd from its bell;
That they for each other
Were made, we see well.

Gleich und Gleich

Ein Blumenglöckchen
Vom Boden hervor
War früh gesprosset
In lieblichem Flor;
Da kam ein Bienchen
Und naschte fein:—
Die müssen wohl beide
Für einander sein.

JOHANN WOLFGANG VON GOETHE, 1814,
TRANSLATED BY EDGAR ALFRED BOWRING

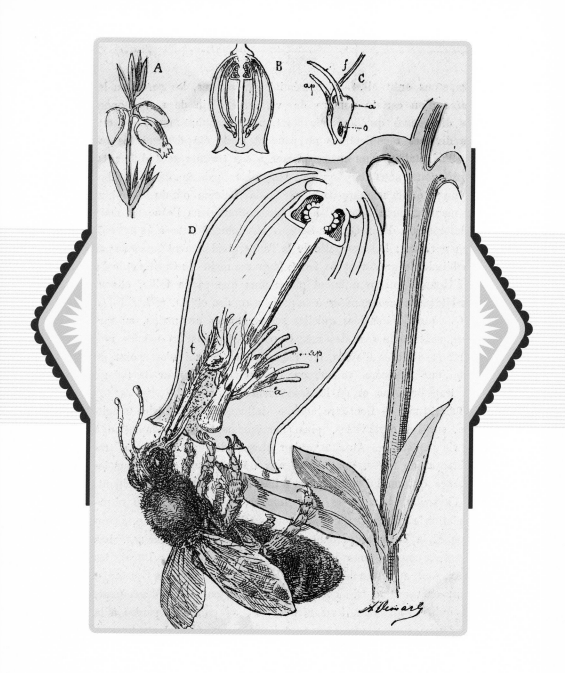

that had been alerted by dancers to a bonanza (a sugar-rich feeder placed in a particular flowerbed) were able to pick out the one and only rewarded patch of flowers among seven hundred kinds of blossoms in a botanical garden.

· · · *Coded Messages* · · ·

And so for several decades, that was how things stood. Foraging bees were believed to inform one another about high-quality sources of food by sharing floral scents, the chemical signature of flowers. Then one day in the mid-1940s, von Frisch and his team more or less idly decided to try something new. Instead of placing feeders close to the hive, as they had done for convenience, they set one of their dishes several hundred yards (about the length of a city block) across country. A simple change, but what a difference! The first thing the researchers noticed was that the number of newcomers searching close to the hive immediately thinned out, but new recruits kept arriving at the distant site. Meanwhile, back home on the dance floor, another surprise was in store. The dancing bees had changed their behavior.

To von Frisch's practiced eye, the difference was striking. Instead of the figure 8 patterns, or "round dances," he had seen when the feeders were nearby, these new "waggle dances" traced out the shape of the Greek letter theta, θ. A dancer still steered to the right and left, but in

> *The round dance, left, and the waggle dance, right, as described by Karl von Frisch.*

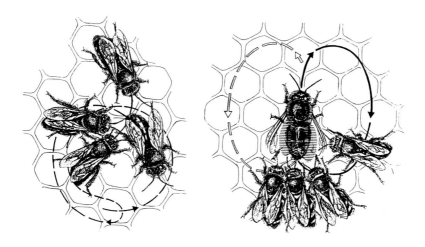

between, she walked straight ahead, shaking her booty from side to side as she moved forward. Could this "waggle run," as von Frisch dubbed the new move, somehow tell other bees that what they were seeking lay at a distance? It would take years of exhaustive investigation by von Frisch, then by von Frisch and Lindauer together, and eventually by their successors to produce a definitive answer to this question. The answer, incredibly, has turned out to be yes.

Bees are bilingual. Not only do they speak the sensory language of the flowers, it has also become clear that they are capable of abstract communication. In experiment after experiment, the length of the waggle phase—and, perhaps more important, of an audible, warm buzz that emanates from the dancer's vibrating body—has turned out to be directly proportionate to the length of the outward-bound journey. On average, eighty milliseconds of waggling represents about one hundred yards of travel, up to a range of nine or ten miles. The longer the journey, the greater the length of the run. Even "round dances" contain a quick waggle or two, an encoded indication that the reward is near.

· · · *The Sun Compass* · · ·

If the fact that honeybees can represent distance had been von Frisch's only discovery, perhaps Lindauer would not have found him in such a state of elation: at his desk before dawn every morning, in the beeyard by 8 AM. But there was more. "I remember the hour," von Frisch recollected later. "It was midday on 15 June 1945 when I realized that all marked dancers which had collected their sugar water at a feeding place about 400 yards due north of the hive performed their wagging run straight down on the vertical honeycomb, whereas unmarked bees which were collecting at other, unknown sites danced in all possible directions." As the hours passed and the sun moved gradually to the west, the marked bees inside the dark hive gradually shifted the angle of their dance.

They say that chance only favors the prepared mind, and von Frisch had been prepared for this observation by recent reports about insect navigation. Researchers had shown that if ants were placed in a level, open environment, with no landmarks to guide them, they could use the sun for reckoning. Was it possible that honeybees on their way to a feeder also oriented themselves to the sun and later incorporated this information into their dances? Dozens of experiments later, von Frisch and Lindauer confirmed that the answer again was affirmative.

Thus, the bees von Frisch had noticed at noon on that fateful June day, with Old Sol high in the southern sky, had oriented their waggle runs straight down the honeycomb, indicating a journey away from the sun, or to the north. Then, as the solar orb slid toward the western horizon and the angle between the feeder and the sun changed, the dancers had adjusted the angle of their dances in smooth progression. By the time the feeder was 120 degrees to the right of the sun, for example, the dancers were slanting their waggle runs 120 degrees to the right of

‹ *A young girl and her flowers are surrounded by the warm buzzing of bees in this nineteenth-century drawing by M. Ellen Edwards.*

vertical. "Against all expectation and probability," von Frisch concluded, the dancing bees "had a 'word' in their 'language' " for direction, as well as for distance.

More remarkable yet, bees that attended the dancers—swirling this way and that in the humming turmoil of the hive—picked up this information, decoded it, and put it into action. If the dancers indicated a feeder to the west, a cloud of new recruits soon headed for it, ignoring identical feeders with identical scents in other directions. And if, the next day, the feeder was moved and the dances reoriented, the dance-followers responded eagerly to their new instructions. The only time the message didn't get through was when the researchers purposely messed things up by flipping a frame of honeycomb out of its usual vertical position and laying it flat. If this horizontal comb was placed in complete darkness so that the bees were deprived of both gravitational orientation and all celestial cues, the dancers performed their waggle runs in random directions. Without meaningful dances, the dance-followers did not know where to go, and very few of them ever found the hidden treasure. But once things returned to normal and the dances made sense again, dozens of new recruits landed directly on target.

And to think that a few decades earlier, von Frisch had found it necessary to argue, against ingrained scepticism, that bees had a capacity for color vision. Now it turned out these marvelous six-legged creatures were also capable of communicating through symbolic movements. Even to him, his findings sometimes seemed "altogether too fantastic" to credit. What was a person to make of experiments in which

A BEE'S
EXPERIENCE

Like trains of cars on tracks of plush
I hear the level bee:
A jar across the flowers goes,
Their velvet masonry

Withstands until the sweet assault
Their chivalry consumes,
While he, victorious, tilts away
To vanquish other blooms.

His feet are shod with gauze,
His helmet is of gold;
His breast, a single onyx
With chrysoprase, inlaid.

His labor is a chant,
His idleness a tune;
Oh, for a bee's experience
Of clovers and of noon!

EMILY DICKINSON, **COMPLETE POEMS**, 1924

foragers were forced to fly out and around a huge rocky ridge to reach a distant feeder but returned home and danced for the A-to-B, straight-line direction? Or what about the times when a dance-follower could be seen checking out one performance after another until she picked up news of a particular patch of flowers where she had fed before, and only then would she head for the door? "No competent scientist *ought* to believe these things on first hearing," von Frisch counseled.

Although his findings were initially published in obscure German-language journals, news quickly spread and soon several respected scientists—notably W. H. Thorpe at Cambridge and Donald R. Griffin at Cornell—had set up temporary apiaries and repeated key experiments. (First, train bees to come to a feeder. Next, bump up the concentration of sugar until the foragers dance for it. Finally, observe as a stream of dance-following recruits locates the experimental feeder among an array of enticing options.) "This memorable experience," Thorpe reported in the prestigious journal *Nature*, "enabled me to resolve to my own satisfaction some of those doubts and difficulties that come to mind on first reading the work, and convinced me of the soundness of the conclusions as a whole."

In the spring of 1949, von Frisch was invited to speak about his dancing bees at scientific institutions across the United States, the first step on his path to the Nobel laureate some twenty-odd years later. Even today, his discovery of the bee's dance "language" is often held to be the most important contribution to the study of animal behavior in a century.

· · · *Flower Power* · · ·

All the same, the path to the Nobel dais was not entirely smooth. In the late 1960s, an American biologist named Adrian Wenner, a one-time-believer-turned skeptic, published a critique of von Frisch's work. Although he agreed that bees perform symbolic dances, he was not convinced that dance-followers put the information to use. ("How much can a honey bee do?" he asked pointedly.) Instead, he hearkened back to von Frisch's early conclusions and revived the idea that bees search for food by relying on floral scents for guidance. As a result of Wenner's objections, researchers set to work designing ever-more-ingenious tests that, in the end, have discredited his major arguments. In 2005, for example, British researchers captured dance-followers as they left the hive, attached miniature transponders to them, and used harmonic radar to track their movements. Blip by blip, the recruits progressed directly toward the target, on the course the dancers had set. Once at

< *Bee orchids do not just rely on scents to attract*
pollinators. Sexy even by floral standards, they draw in male bees
by mimicking the appearance of fresh young queens.

close range—only then—they began to fly this way and that, presumably attempting to locate the feeder by its scent.

Von Frisch wasn't around for this most recent vindication (he died in his beloved Munich in 1982) though had he been, he would likely have relished its clarity. But really, he might have wondered, was more proof necessary? To him, the notion that bee dances were mere nervous spasms didn't make any sense. "How could such a differentiated dance have evolved," he once asked, an uncharacteristic note of exasperation creeping into his tone, "if it were of no significance? Apparently Wenner and his followers don't trouble themselves with this." At some point in the bees' long evolution, either now or in times past, the ability to share information about quality sources of food must have been adaptive. If bees had special powers, it could only be because they needed them.

A honeybee colony is a hungry place, with thousands of mouths to feed. Yet only a relatively small percentage of the work force (about one-quarter of the adult population on any given summer day) is typically engaged in food gathering. All the other members of the colony are either unable to feed themselves—the larvae, the drones, and the queen—or else are young inside workers, engaged in their own duties. Not only must the foragers support these countless dependents during good times, when the flowers are in bloom, but they must also stockpile

> *Honeybee foragers dance to communicate the location of all the resources the colony collects, including propolis (plant resin) and pollen, which are carried home in special structures on the hind legs.*

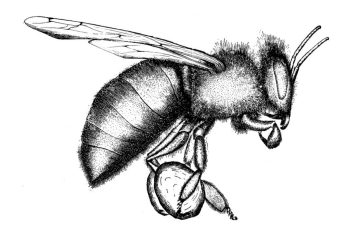

enough resources to sustain the colony through the winter. Then, clustered drowsily inside the hive, the bees will sip on their stored honey and, using its sugars for fuel, keep the hive warm by vibrating their flight muscles. The operation runs year-round on flower power.

Every year, a typical hive requires something like 40 pounds of pollen and 250 pounds of nectar, the latter of which, when mixed with glandular enzymes and thickened, amounts to about 130 pounds of honey. Consider that 1 pound of honey represents the sweetness of about 10 million blossoms. That one forager produces only one twelfth of a teaspoon of honey in her entire career. Consider the miles flown, the lives spent, the prodigious efforts of these insects. And remember, too, that flowers are notoriously fickle. A patch that was in full bloom today, producing copious nectar or pollen, may have faded away by tomorrow. Blossoms that were barren in the morning may be flowing with sweetness by noon. It's as if the countryside beckoned: "Come here, no here, no over here..."

Some bees, a small percentage of the foraging work force, are inclined by their genetics to respond to this blandishment, and out they go, flying from flower to flower to see what they can discover. But when one

THOREAU'S
BEES

THE RAMBLER in the most remote woods and pastures little thinks that the bees which are humming so industriously on the rare wild flowers he is plucking for his herbarium, in some out of the way nook—are like himself ramblers from the village perhaps from his own yard—come to get their honey for his hives—All the honey bees we saw were on the blue-stemmed golden-rod... (not on asters) which emitted a sweet agreeable fragrance—I feel the richer for this experience. It taught me that even the insects in my path are not loafers but have their special errands—not merely & vaguely in this world but in this hour each is about its business. If then there are any sweet sweet flowers still lingering on the hill side it is known to the bees both of the forest & the village. The botanist should make interest with the bees if he would know when the flowers open & when they close.

HENRY DAVID THOREAU,
JOURNAL ENTRY FOR SEPTEMBER 30, 1852

of these scouts strikes it lucky and returns home to dance, her signals trigger a rapid deployment of workers to bring in the harvest. The richer the resource, the more waggle runs the dancer makes to advertise her find, and the more new recruits she ultimately deploys. Thus, a honeybee colony can not only track the ebb and flow of resources over a large area (perhaps forty square miles in all) but also direct its foragers to the best sites at the optimal moment. Biologists speculate that this ability may be especially critical in the tropics, where honeybees rely on large flowering trees that are often miles apart—the very conditions in which *Apis mellifera* is thought to have evolved.

· · · *The Case of the Reluctant Dancers* · · ·

Natural selection is like an eccentric inventor who takes whatever materials are at hand, in the form of DNA, and uses them to work up gadgets, or adaptations, to meet a pressing need. Dance "language," with all its marvels, is an improvised response to the problems of surviving in a world of dispersed resources and time-limited offers. Yet that is not the only challenge that foraging honeybees face. At the same time as they track the supply of essential commodities, they must also remain sensitive to the demand. And that is where Martin Lindauer comes back into the picture.

Like von Frisch before him, Lindauer made a habit of noticing things that other observers missed. He'd catch sight of something in passing— some strange little anomaly or surprising behavioral quirk—and lo and behold, it would turn out to be important. He was particularly good at picking up hints of the subtle connections that bound bee to bee to bee in the tumult of the colony. So trust him to be the first to intuit that the dancers, for all their awareness of distance, direction, and other factors out there in the big wide world, were also remarkably responsive to what was happening back at home, among the inside workers.

It was while he was conducting routine experiments on dance behavior that he picked up the first clues. Why, he wondered, was it sometimes so hard to get his bees to dance for him? On days when the weather was chilly and the deliveries of nectar were low, the foragers would dance for the slightest taste of sugar. But when the sun came out and the nectar flow to the hive increased, the foragers would only respond if his feeders were super-sweet. Apparently, the bees' assessment of the food's value— as indicated by their willingness to dance and advertise it—depended not just on its quality but also on how badly the colony needed it.

Yet how would a forager, working herself to a frazzle flying across hill and dale, keep herself abreast of the colony's changing needs? On the one hand, she might pull out her pocket calculator, tote up the day's deliveries, factor in the energy needed for building comb, divide by the number of larvae in development, push a button, and come up with a daily weighting factor. On the other hand, she was just an insect, with an insect's brain. So Lindauer again took out his stopwatch and settled down to observe what happened when a forager returned to the hive with a load of nectar or water. When the demand for the commodity was high, the incoming bee was beset by eager hive bees, two or three at once, and

> *Many hands make light work for this cheerful cartoon forager.*

her cargo was unloaded in a matter of seconds. When the need was low, by contrast, the poor little beast of burden might have to wander around for a minute, or even two, before finding a single bee to disencumber her.

Lindauer immediately suspected that it was some aspect of this unloading experience— whether the turnaround time or the number of receivers, he was not sure—that reset the nervous system of the forager and recalibrated her dance threshold. This shrewd hypothesis has since been proven correct. If unloaders are removed from a colony so that returning foragers have to wait, then dancing (that is, the recruitment of additional foragers) is suppressed, even if the colony is facing starvation. Under normal circumstances, however, the unloading experience provides foragers with an accurate readout of the colony's current needs and permits them to tune their behavior accordingly.

As it happens, this "unloading-reset button" also holds the key to another unsolved mystery. In the Case of the Overheated Colony, you'll remember, a spike in the temperature inside the hive had somehow motivated outside workers to increase their delivery of water. Lindauer was able to show that here too, the connection was made face to face, when foragers returned to the hive for unloading. Any bee that showed up with a droplet of water received a tumultuous welcome and was inspired to

dance, thereby recruiting more foragers to deliver coolant. It all seems so simple and elegant.

But don't be misled. According to recent findings by Thomas Seeley, a lifelong honeybee enthusiast and now a professor at Cornell, communication between inside and outside workers is actually a quirky and complicated business. Let's say, for example, that a forager that has discovered a highly profitable nectar source finds herself waiting at the unloading dock, time ticking past, for more than about forty seconds. Rather than give up quietly, she may decide to demand better service by quavering slowly around the comb, all a-tremble. Several times a minute, she lunges forward, beeps loudly and butts into another bee, before continuing on her erratic journey. This "tremble dance" serves as a signal for any waggle-dancers she bumps into to stop recruiting. At the same time, her performance also stimulates hive bees to rush to the unloading dock and reduce the wait times by getting to work.

The governance of a honeybee colony has turned out to be more extraordinary than anyone, even von Frisch in his most elated moments, could have imagined. Through the simplest of physical means—a waggling abdomen, a pause in the action, a quivering run—thousands of individual bees pool their awareness of both the world around the hive and the world within it, to achieve near-perfect coordination. Consider the honeybee: an insect.

‹ *An Asian honeybee,* Apis cerana, *is blurred by the movements of her "cleaning dance." Stimulated by this behavior, her companions will groom her body.*

4

LIFE

LESSONS

Go to the bee,

then poet,

consider her ways

and be wise.

GEORGE BERNARD SHAW, 1856–1950

When Martin Lindauer had embarked on his research all those years ago, the world had been aflame with hatred. Ordinary people—good people who under other circumstances might never have hurt a soul—had succumbed to the seductions of demagogues. It seemed that there was a tragic gap in human affairs between the individual and the crowd, between the common decency of neighbors and the brutality of the mob. And so it had been a relief and a comforting distraction for Lindauer to spend his time among creatures that were adapted to serve the common good and to share their knowledge for the benefit of the group. For honeybees, cooperation was not an impossible dream to be pursued; it was the day-to-day foundation of survival.

Like anthropologists studying an exotic human culture, Lindauer and his colleagues had helped to uncover some of the principal means through which honeybees achieve this enviable harmony. By observing Bee 107, for example, he had confirmed that the abilities of a worker bee depend, in large part, on how old she is and that this age-based developmental sequence provides the basis for the colony's division of labor. More important, Bee 107 had also demonstrated that individual

bees, whatever their ages or stages, are unexpectedly capable and alert, equipped to sense and respond to events around them. And then, to top it all off, had come the discovery of the foragers' dance "language," with its exquisite sophistication, and the bees' unparalleled powers of communication. A honeybee colony had turned out to be kind of fabulous living gizmo—a composite being—in which thousands of closely related individuals, each highly competent in itself, were connected by a network of shared experience.

⌃ Bound together by their mutual attraction to their mother-queen, a colony of bees allows itself to be collected from a temporary resting place and housed in a new home.

Much as a multitude of cells make up our bodies, so a multitude of bees make up the super-organism of the colony. And it was thanks to this mind-bending insight that Lindauer embarked on what he would later refer to as "the most beautiful experience" of his life. It had all begun on a spring day in the late 1940s, when he was walking through the apiary where he conducted his research and happened upon a large, beard-shaped cluster of honeybees hanging on a bush. In itself, this was not surprising, since beekeepers had long observed, and often lamented, the tendency of colonies to swarm, particularly if their hives become overcrowded. When this happens, about a third of the population typically stays home and rears a new queen, while the balance—a party of perhaps ten thousand, including workers of all ages—rush off with the old monarch, their mother. These emigrants settle in a cluster and literally hang out together for anywhere from a few hours to several days. Then (unless a beekeeper comes to collect them, by shaking them into a box) they take off en masse and fly directly to a new nest site, often in a hollow tree. In this way, the original hive produces a daughter colony.

· · · *Dirty Dancers* · · ·

How do honeybees accomplish this amazing feat? Lindauer didn't know, but as he watched the golden mass of humming insects, he did notice something odd. A few of the bees on the surface of the swarm were waggle dancing. At first he assumed they must be ordinary foragers advertising sources of food—but why did so many of them look grubby? When he brushed debris from their dusty bodies and put it under a microscope,

ANGEL OF BEES

The honeycomb
that is the mind
storing things

crammed with sweetness,
eggs about to hatch—
the slow thoughts

growing wings and legs,
humming memory's
five seasons, dancing

in the brain's blue light,
each turn and tumble
full of consequence,

distance and desire.
Dangerous to disturb
this hive, inventing clover.

How the mind wants
to be free of you,
move with the swarm,

ascend in the shape
of a blossoming tree—
your head on the pillow

emptied of scent and colour,
winter's cold indifference
moving in.

LORNA CROZIER, 1992

he never found grains of pollen, as he expected to. Instead, these bees were just plain dirty: "Black with soot," he reported, "red with tile-dust, white with flour, or gray and dusty as if they had been grubbing in holes in the ground."

There was only one possible explanation, however fantastic it seemed. With the swarm hanging in limbo, these dancers must have been out searching for nest sites. An enchanting hypothesis, but how to prove it?

By the time Lindauer seriously took up this challenge, he had returned to Munich, where, even years after the war, there was no shortage of ruined buildings, abandoned attics with forgotten flour chests, or disused chimneys for the bees to discover and explore. And so it was that the good burghers of München were treated to the spectacle of Lindauer and his assistants rushing along streets and down alleys, ordnance maps in hand, searching for bees with daubs of paint on them. The plan was to mark the dancing bees on the swarm, decode their dances—a stopwatch to measure distance, a compass for direction—map their destinations, and then head out and try to observe them in action. Sure enough, when the bees were relocated (as they occasionally were), they were never found in or around flowers. Instead, they could be seen crawling into dusty knotholes, inspecting cracks in windowsills, or flying in and out of abandoned woodpecker nests. Lindauer's dirty dancers really were house hunters.

‹ *Bivouacked in a pine tree, a swarm of several*
thousand bees hangs in a peaceful, pendulous company.

· · · *Location, Location, Location* · · ·

Meanwhile, back at the swarm, there was more to learn. If one sat patiently and watched the dancing bees for hours on end, it became clear that they were not all advertising the same location. In one typical case, the day opened with the announcement of a single site three hundred yards to the south, but soon there came news of another almost a mile away in another direction. By nightfall, six more sites were in contention, and the next day brought fourteen more, for a grand total of twenty-two. Yet the swarm, with its lone queen, could not afford to split up. To survive, the bees had to settle in a common nest, and the sooner they did so, the better. Now the mystery took on a sharper focus: how could a group of insects, hanging in a bush, make such a complex decision?

In the end, answering this question would take the combined efforts of Lindauer and his scientific successors—notably, the impresario of the tremble dance, Thomas Seeley. Thanks to them, we know that each swarm typically fields a few hundred house hunters (about 5 percent of the population), all of which have previous experience as foragers. Now radically reconfigured to search for dark crevices instead of bright blossoms, these intrepid explorers set out to scour the countryside. When one of them locates a potential property, she stops and examines it, both by walking up, down, and around the inner walls of the cavity and by making brief excursions outdoors, a process that often takes her half an hour or more.

> ➤ *There is more than one way to catch a swarm of bees, as this jaunty French* apiculteur *demonstrates.*

    The ultimate dream home for honeybees is an enclosed hollow with thick, leak-proof walls, a south-facing entrance close to the floor, and a total living space of around ten gallons (about the size of a small aquarium). As the scout bee conducts her inspection, she checks on these and other features—the height and size of the entranceway, for example, and the presence or absence of drafts—and obtains a sense of the property's value. Find this hard to believe? Seeley did as well, so he decided to test the bees' abilities by playing a trick on them. To do so, he built a special nest box in which the inner walls could be rotated forward and back

relative to the door, which stayed fixed in position. Thus, if a scout came in through the entrance as the walls were rotating forward, she would be carried along and quickly find herself back where she had begun. But if the walls were turning against her, she would have to take many more steps to make her way around to where she had started.

After running dozens of bees on this treadmill, Seeley noted that the ones on the "short" track were unlikely to dance, presumably because their feet told them that the box was too small to be adequate. But those that had to walk a "long" distance judged the box to be large enough, and they recommended it to their sisters with enthusiasm. Every step was a measurement.

· · · *Forming a Quorum* · · ·

Once a site has passed inspection, the scout that makes the discovery returns to the swarm and dances, thereby sharing her findings with other, as-yet-uncommitted house hunters. The more highly she values her property, the more waggle runs she performs, and the more whirling, twirling followers she attracts. Each of these new recruits then flies out, locates the site for herself, and makes an independent evaluation. If the proposed accommodation fails this renewed scrutiny, the recruit will not dance for it and the error of the original scout is soon silenced. But if the property meets her standards, the recruit picks up the beat and joins in the excitement of the marathon dance party.

The scene seems ripe for riot, as dancers crowd onto the floor, all more or less excited and all badly informed. As a rule, each scout knows only about one of the options in contention—the one she is dancing for—and therefore is unable to compare it to any of the others. Yet out of this incipient chaos, order begins to emerge. For at the same time as new recruits are drawn into the action, other dancers start dropping out, beginning with those that discovered low-quality accommodations. As a result, the fixer-uppers quickly fall out of contention, and the up-market options—the places for which the dancing has always been most intense—begin to dominate the discussion. Soon, the best site

< *In the Middle Ages, it was believed that "tanging," or creating a loud clanging racket, would cause a swarm of bees to settle into a hive.*

pulls ahead of the others and becomes the unanimous choice, as all the remaining dancers step into line.

In fact, even before this happens, the decision has been made. Through a series of ingenious manipulations, Seeley has shown that the "tipping point" in the bees' debate comes not with complete consensus but sometime earlier, just as the tide of opinion is beginning to turn. Why wait for full agreement if it is possible to sense the outcome in advance and move on to the next step? Remember how the unloading experience of a forager provided her with an instant readout of the colony's needs? In much the same way, the experience of a house hunter provides her with a summary of the colony's decision making. If a bee goes out on a site inspection and finds herself alone, the absence of other bees speaks volumes. But if a dozen or more other house hunters are inspecting the cavity along with her (a signal that ten times that number are interested in the place), her nervous system flashes "Yes!"

Once a scout senses that a choice has been made, she turns into a rabble-rouser. Her new mission is to animate the resting swarm (the 95 percent of the population that don't know anything about what has been going on) and prepare it for the flight to its new home. Like a tremble dancer on steroids, she starts running erratically over and through the cluster, ramming into other bees and revving her wings. Not content

> ➤ *Once a colony settles into a new nest, the bees hook together, toe to toe, as shown on the left of this illustration. Hanging quietly in the darkness, they secrete wax and furnish their home with sheets of honeycomb.*

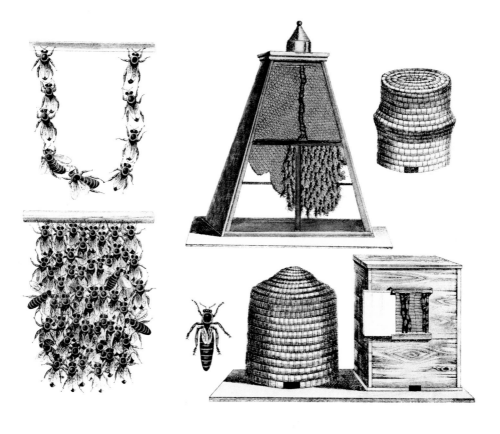

with mere persuasion, she occasionally grabs hold of a drowsy swarm-mate, presses her thorax to it, and hits it with a sharp blast of vibration. "*Vroom:* Warm up your flight muscles. Ready, set, go." As more and more buzz-runners go into action, the pulsing hum emanating from the swarm rises to a crescendo. Then suddenly, the solid surface of the cluster disintegrates and the entire swarm takes wing, filling the air with the roar of ten thousand insects. Apparently guided by scouts that streak rapidly through the cloud of bees and show it where to go, the colony sweeps to its chosen home.

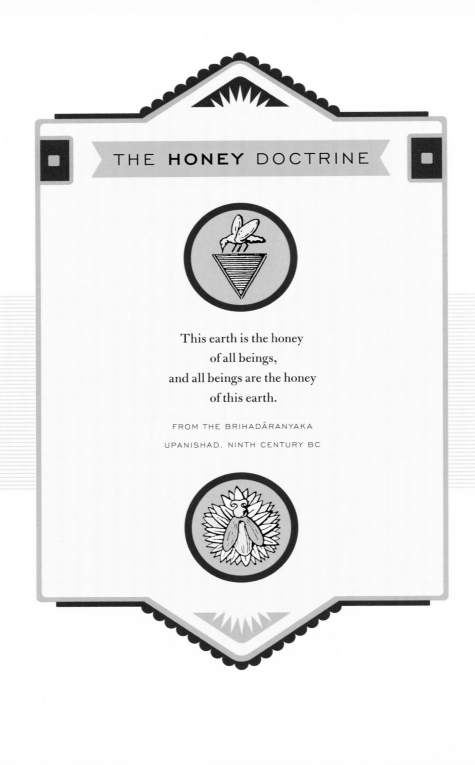

THE **HONEY** DOCTRINE

This earth is the honey
of all beings,
and all beings are the honey
of this earth.

FROM THE BRIHADÂRANYAKA
UPANISHAD, NINTH CENTURY BC

· · · *Nurturing Your Inner Insect* · · ·

In 1959, exactly a decade after von Frisch and his dancing bees had
made their triumphant tour of the United States, Lindauer found himself
following in his mentor's footsteps. Despite his protestations that, at 41,
he was much too young by German standards for international renown,
the house-hunting bees had established his reputation. And so that
spring, a man who had begun his career in a half-ruined garden, alone
with his bees, found himself speaking to large and admiring audiences
at North American universities. Some months later his name appeared
on the cover of a slender volume, *Communication Among Social Bees,*
which was based on lectures he had delivered at Harvard. In the preface
to this publication, he expressed his relief at discovering how "common
problems and interests within biology [had] built new bridges... that
span wrongs of the past." His bees had helped to renew the possibility of
friendship between nations.

Bees had brought sweetness out of chaos. Could it be that they really
do have something to teach us about living in harmony? For Lindauer,
this was merely a pleasing notion, but for his successors, it has become a
subject for serious, straight-on consideration. In a recent paper on group
decision making by house-hunting swarms, for example, Seeley and his
co-workers reflect on the bees' almost faultless ability, working as a com-
mittee of the whole, to choose the best nest site on offer: "It is clear they
are successful at making collective judgments." Unlike human groups,
which often seem less intelligent than the individuals who make them up,
a swarm of bees is always smarter than the sum of its parts. So what is it

about the bees' group process that boosts their collective IQ and permits them to avoid costly errors? Could their strategies possibly take the sting out of our own decision making?

Employing insects as management gurus has its limitations, and no one recommends that we blindly attempt to imitate them. (In particular, running around the conference room and accosting colleagues, channeling your inner bee, is likely to be frowned upon in most settings.) All the same, Seeley and his colleagues have identified three basic characteristics of a bee swarm that, they believe, contribute to its exceptional abilities. First, the house-hunting process is open to the widest possible input of knowledge and ideas. Rather than being "led or dominated by a small number of bees," the researchers note, "the decision-making process is broadly diffused among all the [scouts] in a swarm." Second, each individual that contributes to the debate has the courage of her convictions and makes her own independent assessment of the situation. ("She does not blindly support the bee whose dance she followed.") And finally, the process is structured not as a search for compromise but as a friendly contest. Debate continues unhindered until the opinions of the participants begin to coalesce, thereby ensuring that all the competing options are fully assessed. ("The quorum-sensing method of aggregating the bees' information allows diversity of opinion to thrive, but only long enough to ensure that a decision error is improbable.")

> *Who would ever have guessed that a bunch of insects*
hanging in a tree is actually a kind of composite thinking machine?

Pl. 37. pag. 668. Mem. 12. de l'Hist. des Insectes Tom. 6.

Fig. 2.

Fig. 1.

Although these Three Habits of Highly Intelligent Groups have not yet been applied on the global stage, Seeley has employed them, with success, in a university department. If they help bring peace to that notorious hornets' nest, they must have something going for them.

· · · *Deep Thinkers* · · ·

All in all, the idea that we can learn something about ourselves by studying bees is a surprisingly hot property in contemporary science. At the same time as Seeley is engaging his colleagues at Cornell in the hubbub of frank debate, researchers at other institutions are looking to bees for

insights into human nature. In Germany, France, and Australia, for example, researchers (chief among them Randolf Menzel, one of Lindauer's former students) are studying the inner workings of the honeybees' tiny brain, an organ the size of a fly speck that nonetheless acquires information, remembers it, and decodes symbolic messages. What's more, recent research by Menzel and his colleagues demonstrates that the bees' "amazing mini-brain" is even capable of framing concepts.

Imagine an experimental setup in which bees—think flying lab rats—are trained to enter a Y-shaped tube, one arm of which is rewarded with sugar water. As the foragers zoom in through the entrance, they are presented with a swath of color (yellow, say, or blue) or a whiff of scent (it might be mango or lemon). At the fork of the Y, they are again presented with a stimulus—either like or unlike the original clue—that directs them down the right path to find food. Once the bees have learned the rules of the game, by matching sample to sample or looking for difference, the researchers intervene to make things more complicated. Now, instead of receiving a clear signal at the junction, the bees are presented with options and required to choose between them. Not only do the bees rise to the challenge by following the rules they have learned, but they are also able to apply their knowledge to new situations. Thus, if bees that were trained with colors are retested with scents, they respond correctly by detecting the abstract quality of "sameness" or "difference."

· · · *Genomic Virtuosi* · · ·

The brain of a honeybee consists of fewer than a million neurons, or less than .00001 percent as many as our own. Yet given its remarkable abilities, researchers believe that studying this relatively simple system may help us understand a little more about what is going on inside our own noggins. A similar hope is voiced by some of the more optimistic and daring members of an international research consortium that, in 2006, sequenced the genome of the honeybee. Only the fifth species of insect to be accorded this attention (after two fruit flies, the malaria mosquito, and the silk moth), honeybees were singled out for study partly for practical reasons—their importance to pollination—but also because, frankly, we're besotted with them. The more we learn about them, the more we yearn to understand what makes them such bright little sparks of wonderment.

The human genome is currently estimated to contain between 20,000 and 25,000 genes, far fewer than was originally predicted and only one

‹ *The human-to-honeybee connection*
is both ancient and completely contemporary.

notch above the 19,000 of a silk moth. The honeybee has also come in below expectations, with a preliminary total of about 10,000. (Even the humble *Drosophila* scores higher.) But it seems that this is a case in which the size of one's natural endowment really does matter less than what one is able to do with it. What counts, the experts tell us, is not just the number of bells and whistles in an organism's genetic program but the finesse with which those special features can be switched on and off, or dialed up and down, in response to changing conditions. At the most basic level, the remarkable flexibility of honeybee behavior appears to result from this process of "gene regulation," a suite of biochemical pathways that translates the events of everyday experience—the thump of a buzz-runner, for instance, or the sight of a blue blossom—into messenger molecules that genes can detect and respond to.

Given the opportunity, you could catch a glimpse of this interaction at first hand by replicating a slightly gruesome experiment. It involves capturing young nurse bees, whirring their brains in a blender, and running the resulting juices through a microassay system, to see which genes are activated. If you then repeat the process with forager bees (rather than nurses), you will observe that a strikingly different set of genes is at work. Although this line of research is still in its infancy, it is already challenging our old image of genetics as a rigid blueprint. Instead, we may soon find ourselves thinking of the genome as a kind of instrument,

➤ *Unlike the honeybee, most of the world's diverse species of wild bees have scarcely been studied—an untapped store of wonder.*

a spiral keyboard perhaps, that is played upon by life experience. The virtuosi of gene regulation—the species in which the genome is most responsive to environmental input—appear to be sophisticated, social species, like honeybees and humans.

If this all sounds a little esoteric, then consider the views of the happily named Gene Robinson of the University of Illinois, a one-time-beekeeper-turned-high-end-researcher and the co-leader of the genome-sequencing project. For him, demonstrating the link between a worker bee's genes and her behavior, through the brains-in-the-blender experiments, is of deep personal significance. As the child of Holocaust survivors, he often thinks of the darkness of the still-recent past when a grim science of genetic determinism filled the death camps. Given this background, he feels a special responsibility to shine a light on what honeybees have to teach us about the dynamic interaction between genetics and experience. "We are developing a new framework for understanding the relationship between nature and nurture," he says, and an inducement for human tolerance.

When the last bee died,
Nobody noticed. Nobody put on black
Or made a dirge for the death
Of honey. Nobody wrote an elegy
To apricots, no one mourned for cherries.

When the last bee died,
Everyone was busy. They had things to do,
Drove straight to work each morning,
Straight back home each night. The roads
All seriously hummed. Besides,

The pantries were still packed
With cans of fruit cocktail in heavy syrup,
Deep deep freezers full
Of concentrated grape and orange juice,
Stores stocked with artificial flavoring.

When the last bee died, nobody saw
The poppies winking out, nobody cried
For burdock, yarrow, wild delphinium.
Now and again a child would ask for
Dandelions, quickly shushed: That pest!

And everyone is fine. The children healthy,
Radish-cheeked. They play she loves me/not
With Savoy cabbage leaves, enjoy the telling
Of the great myths, peach and peony.
No one believes in apples any more.

BETTY LIES, 1998

· · · *Coda: Silent Spring* · · ·

Just when we seem to be on the verge of resolving some fundamental questions about both bees and ourselves, we wake up one morning to media stories, emanating from the United States, that thousands of honeybee colonies have vanished. The trouble first surfaced in the fall of 2006, when a commercial beekeeper in Florida discovered that hive after hive was empty or, at best, occupied only by the queen and a few attendants. The rest of the bees had vanished without a trace, not even leaving behind their dead bodies. Soon, similar stories were coming in from others parts of the United States, together with reports of unusually high winter die-outs from severe weather and other causes. As many as 875,000 bee colonies are estimated to have been lost in the fall and winter of 2006–07, or about 32 percent of America's managed bee population. Since then, overwinter losses of over 30 percent have become the new normal for beekeepers in the U.S. In Canada, where the statistics have also been grim, there was a welcome uptick in 2009–10, with losses approaching the long-term average of about 15 percent.

This is not the first time that commercially managed honeybees have suffered a severe hit. Over the years, the vanishing-bee syndrome, or something resembling it, has also struck in the United Kingdom, Australia, Canada, and France. In each of these instances, the crises were comparatively short-lived and bee operations eventually rebounded. What distinguishes the current situation is not so much the severity of

> ➤ *In the Middle Ages, the only hazards that a beekeeper had to fear were sweet-toothed intruders and unfavorable weather.*

the losses as their duration, their continent-wide extent and the heightened public anxiety that surrounds them. For though lacking in dead bodies, the modern, industrial apiary is strewn with smoking guns, and virtually all of them implicate us.

We live in a world awash in pesticides, overrun by invasive pests, afflicted by the introduction of exotic diseases. All of these factors and others, alone and in combination, are probable causes of the still-unexplained losses. So too is semi-starvation. Many of the colonies that vanished in 2006–07 had spent the spring and summer piled onto flatbeds, traveling the continent as an itinerant pollination brigade. All told, commercial honeybees are the primary pollinators for no fewer than 90 fruit and vegetable crops, which together make up the most nutritious third of our daily diet. But in the process of nourishing us, the bees are likely to become ill-nourished themselves, because they are forced to draw their sustenance from one single-crop landscape after another. Perhaps the stress of life in the fast lane has left them

susceptible to mites or infections or poisons, or all of the above, on a downward spiral to silence.

The stillness of an abandoned beeyard echoes eerily through our thoughts, awakening our worst fears about our relationship with the natural world. What would become of us, and the living beauty of the Earth, without the buzz of pollinators? And it is not just honeybees that demand our attention but also their wild cousins, the whole busy tribe of diggers, masons, carpenters, leaf-cutters, orchard bees, bumblebees, and all the rest, that contribute their services, gratis, in every garden, field, and forest around the planet. They are out there in all their diversity, helping to keep the world flowering and fruitful. Some of them are known to be declining in numbers, and several are feared to have gone extinct. According to the Xerces Society for Invertebrate Conservation, at least fifty-eight species of North American bees are currently at risk of vanishing from the planet. Meanwhile, researchers in northern Europe have recently documented widespread declines of both wild bees and bee-pollinated plants in the Netherlands and England.

What bees ask of us is simple: a world free from poisons and other stressors, with places where they can nest and a sweet, season-long supply of flowering plants. In return, they offer to teach us their deepest lesson yet. Much as a honeybee belongs to her colony, so we humans belong to the living community of the Earth. The wild lies all around us, and we draw it in like breath. Our lives are indivisible from the lives of insects.

< *Native plants, like this wild sunflower, provide*
abundant resources for native bees, including this shining little beauty.

ACKNOWLEDGMENTS

In undertaking this project, I have been sustained by the expertise and enthusiasm of many people. They include Gro Amdam of Arizona State University, Bryan N. Danforth and Margaret K. Kirkland of Cornell University, Peter Kevan of the University of Guelph, Faisal Moola of the David Suzuki Foundation, Deborah Smith of the University of Kansas, Sharon Sollosy of T&H Apiaries, Marla Spivak of the University of Minnesota, and Mark Winston of Simon Fraser University. Special thanks are due to Gene E. Robinson of the University of Illinois Urbana-Champaign and to Thomas D. Seeley of Cornell University, both of whom reviewed the text and were otherwise extravagantly generous with their time and knowledge. Finally, it is a pleasure to acknowledge the unfailing kindness of my partner, Keith Bell, the perfect companion on a backyard safari or any other adventure.

NOTES

Notes refer to direct quotations only.

v Charles Butler, "To the Reader," *The Feminine Monarchie, Or A Treatise Concerning Bees and the Divine Ordering of Them,* Oxford, 1609. Available online at http://www.shipbrook.com/jeff/bookshelf/details.html?bookid+16

INTRODUCTION

1 E.O. Wilson, "The Little Things That Run the World (The Importance and Conservation of Invertebrates)," *Conservation Biology* 1 (1987): 344–46.

3 Pliny the Elder, *Natural History,* Book 11, Chapter 17 (London: Taylor and Francis, 1885). Available online at http://www.perseus.tufts.edu

3 Virgil, *The Georgics,* Book IV, line 220. Available online at http://www.tonykline.co.uk

4 E.J. Jenkinson, *The Unwritten Sayings of Jesus,* 1925, as quoted by Hilda M. Ransome, The Sacred Bee in Ancient Times and Folklore (London: George Allen Unwin, 1937), 74.

CHAPTER I

7 Karl von Frisch, as quoted by Martin Lindauer, "Karl von Frisch, a Pioneer in Sensory Physiology and Experimental Sociobiology," in *Neurobiology and Behavior of Honeybees,* ed. Randolf Menzel and Alison Mercer (Berlin: Springer-Verlag, 1987), 6.

9 Karl von Frisch, re "mixed descent," *A Biologist Remembers* (Oxford: Pergamon Press, 1967), 131.

10 Karl von Frisch, re the antiquity and perfection of bees, *A Biologist Remembers,* 106.

11 Katharine Tynan, "Telling the Bees," *Herb o' Grace: Poems in War Time* (London: Sidgwick and Jackson, 1918). Available online at http://www.beck.library.emory.edu/greatwar/poetry/view.php?id=herb_003

14 "Aesop's bees," adapted from *The Aesop for Children* (Chicago: Rand McNally, 1919). Available online at http://www.projectgutenberg.com

23 Ralph Waldo Emerson, "The Humble-
Bee," *Yale Book of American Verse* (New
Haven: Yale University Press, 1912).
Available online at http://www.bartleby.
com/102/

24 Stingless bees not biteless, quoted by E.O.
Wilson, *In Search of Nature* (Washington,
D.C.: Island Press, 1996), 79.

27 William Butler Yeats, "The Lake Isle
of Innisfree," *Modern British Poetry,* ed.
Louis Untermeyer (New York: Harcourt,
Brace and Howe, 1920). Available online
at http://www.bartleby.com/103/

28 Karl von Frisch, re Nobel Prize,
as quoted by Thomas D. Seeley, S.
Kunholz, and R. H. Seeley, "An Early
Chapter in Behavioral Physiology and
Sociobiology: The Science of Martin
Lindauer," *Journal of Comparative
Physiology A* 188 (2002): 452.

28 Karl von Frisch, re demonstration,
A Biologist Remembers, 57.

30 "The god Re," quoted by John B. Free,
Bees and Mankind (London: Allen and
Unwin, 1982), 93.

31 Karl von Frisch, re "command,"
A Biologist Remembers, 57.

31 Karl von Frisch, re intelligence service,
A Biologist Remembers, 71.

CHAPTER 2

33 Charles Butler, *The Feminine Monarchie,
Or A Treatise Concerning Bees and the
Divine Ordering of Them,* Oxford, 1609.

Available online at http://www.shipbrook.
com/jeff/bookshelf/detailshtml?bookid+16

35 Martin Lindauer, re Hitler years,
as quoted by Thomas D. Seeley, S.
Kuhnholz, and R.H. Seeley. "An Early
Chapter in Behavioral Physiology and
Sociobiology: The Science of Martin
Lindauer," *Journal of Comparative
Physiology A* 188 (2002): 445.

37 Martin Lindauer, re confusion in hive,
Communication among Social Bees.
(Cambridge, MA: Harvard University
Press, 1961), 3.

38–39 Pliny the Elder, *Natural History,*
Book 11, Chapter 17 (London: Taylor
and Francis, 1885). Available online
at http://www.perseus.tufts.edu

41 Karl von Frisch, re character of drones,
*Bees: Their Vision, Chemical Senses, and
Language* (Ithaca: Cornell University
Press, 1950), 1.

43 Karl von Frisch, re massacre of drones,
*The Dancing Bees: An Account of the Life
and Senses of the Honey Bee,* tran. Dora
Ilse (New York: Harcourt, Brace, 1953), 32.

46 William Shakespeare, *Henry V* 1(2):
187–204.

50 Karl von Frisch, re the antiquity
and perfection of bees, *A Biologist
Remembers,* 106.

52 Isaac Watts, "Against Idleness and
Mischief," *Divine Songs Attempted in
Easy Language for the Use of Children*
(London: M. Lawrence, 1715). Available

online at http://www.rpo.library.utoronto.
ca/poem/2265.html

54 Martin Lindauer, re activities of Bee 107,
    Communication Among Social Bees, 21.

57 Martin Lindauer, re division of labor,
    Communication Among Social Bees, 21.

CHAPTER 3

59 Virgil, *Aeneid:* 430–36, trans. John
    Dryden, as quoted by James L. Gould
    and Carol Grant Gould, *The Honey Bee*
    (New York: Scientific American Library,
    1988), 1.

60 Karl von Frisch, re disbelief, *A Biologist
    Remembers,* 142.

63 Karl von Frisch, re comet's tail, *The
    Dancing Bees: An Account of the Life and
    Senses of the Honey Bee,* 101, 103.

64 Johann Wolfgang von Goethe, "Like
    and Like," *The Poems of Goethe,* trans.
    Edgar Alfred Bowring (London:
    John B. Alden, 1853). Available online
    at http://www.gutenberg.org/dirs/
    etext98/tpgth10.txt

66 "Newcomers arriving at distant site,"
    paraphrased from Karl von Frisch,
    The Dancing Bees, 116.

68 Karl von Frisch, re 15 June 1945,
    A Biologist Remembers, 143.

70 Karl von Frisch, re "words" for distance
    and direction, *A Biologist Remembers,*
    140, 143.

70 Karl von Frisch, "altogether too fantastic,"
    A Biologist Remembers, 140.

71 Emily Dickinson, "A Bee's Experience,"
    The Complete Poems of Emily Dickinson
    (Boston: Little Brown, 1924).
    Available online at http://www.bartley.
    com/br/113.html

72 Karl von Frisch, re advice to scientists,
    as quoted by Donald R. Griffin,
    "Introduction," *Bees: Their Vision,
    Chemical Senses, and Language,* vii.

72 W.H. Thorpe, as quoted by Donald R.
    Griffin, "Introduction," *Bees,* viii.

73 Adrian M. Wenner, *The Bee Language
    Controversy: An Experience in Science*
    (Boulder: Education Programs Improve-
    ment Corporation, 1971), 26.

74 Karl von Frisch, re evolution of
    differentiated dance, letter to James
    Gould, 1975, quoted by Tania Muntz,
    "The Bee Battles: Karl von Frisch, Adrian
    Wenner and the Bee Dance Language
    Controversy," *Journal of the History of
    Biology* 38 (2005): 559.

76 Henry David Thoreau, *Journal,* Volume
    5: 1852–1853, ed. Patrick F. O'Connell
    (Princeton, NJ: Princeton University
    Press, 1997), 363.

CHAPTER 4

83 George Bernard Shaw, *Man and Super-
    man,* Act II, line 81 (Cambridge, MA: The
    University Press, 1903). Available online
    at http://www.bartleby.com/157/2.html

86 Martin Lindauer, re beautiful
    experience, as quoted by Thomas

D. Seeley, S. Kuhnholz, and R.H. Seeley, "An Early Chapter in Behavioral Physiology and Sociobiology: The Science of Martin Lindauer," *Journal of Comparative Physiology A* 188 (2002): 447.

87 Lorna Crozier, "Angel of Bees," *Inventing the Hawk* (Toronto: McClelland and Stewart, 1992), 51.

89 Martin Lindauer, re dusty bees, "House-Hunting by Honey Bee Swarms," trans. P. Kirk Visscher, Karin Behrne, and Susanne Kuhnholz, *Zeitschrift fur vergleichende Physiologie* 37 (1955): 266.

96 *B̓rihâdaranyaka Upanishad,* Book II, Chapter v, trans. Max Muller, in *Sacred Texts of the East,* Vol. 15 (Oxford: Oxford University Press, 1884). Available online at http://www.sacred-texts/com

97 Martin Lindauer, re building bridges, *Communication Among Social Bees,* v.

98 Thomas D. Seeley, P. Kirk Visscher, and Kevin M. Passino, "Group Decision Making in Honey Bee Swarms," *American Scientist* 94 (2006): 228.

103 Gene E. Robinson, personal communication, 2007.

104–105 Betty Lies, "End Notes for a Small History," *Southern Poetry Review* 38 (Summer 1998): 33.

SELECTED RESOURCES

CHAPTER I

BEE LORE

Bishop, Holley. *Robbing the Bees: A Biography of Honey—the Sweet Liquid Gold That Seduced the World.* New York: Free Press, 2005.

Cook, Arthur Bernard. "The Bee in Greek Mythology." *Journal of Hellenic Studies* 15 (1895): 1–24.

Crane, Eva. *The Archaeology of Beekeeping.* London: Duckworth, 1983.

——. "A Short History of Knowledge About Honey Bees (*Apis*) Up to 1800." *Bee World* 85 (2004), no. 1: 6–11.

——. *The World History of Beekeeping and Honey Hunting.* New York: Routledge, 1999.

Ellis, Hattie. *Sweetness and Light: The Mysterious History of the Honey Bee.* London: Sceptre, 2004.

Fraser, H. Malcolm. *Beekeeping in Antiquity.* London: University of London Press, 1931.

Free, John B. *Bees and Mankind.* London: Allen and Unwin, 1982.

Morley, Margaret Warner. *The Honey-Makers.* Chicago: A.C. McClurg, 1915.

Preston, Claire. *Bee.* London: Reaktion Books, 2006.

Ransome, Hilda M. *The Sacred Bee in Ancient Times and Folklore.* London: George Allen and Unwin, 1937.

KARL VON FRISCH AND

MARTIN LINDAUER: BIOGRAPHY

Lindauer, Martin. "Karl von Frisch, a Pioneer in Sensory Physiology and Experimental Sociobiology." In *Neurobiology and Behavior of Honeybees,* edited by Randolf Menzel and Alison Mercer. Berlin: Springer-Verlag, 1987.

Seeley, Thomas D., S. Kuhnholz, and R.H. Seeley. "An Early Chapter in Behavioral Physiology and Sociobiology: The Science of Martin Lindauer." *Journal of Comparative Physiology A* 188 (2002): 439–53.

von Frisch, Karl. *A Biologist Remembers.* Translated by Lisbeth Gombrich. Oxford: Pergamon Press, 1967.

BEES OF THE WORLD

Danford, Bryan. "Bees." *Current Biology* 17 (2007): 156–61.

Michener, Charles D. *The Bees of the World.* Baltimore: Johns Hopkins University Press, 2007.

O'Toole, Chris. "A Manifesto for Bee Taxonomy—Bee Faunas and the Taxonomic Deficit: Consequences and Solutions." Available online at http://www.fao.org/ag/AGp/AGPS/c-CAB/Castudies/pdf/3-002.pdf

O'Toole, Christopher, and Anthony Raw. *Bees of the World.* New York: Facts on File, 1991.

Poinar, G.O., Jr., and B.N. Danforth. "A Fossil Bee from Early Cretaceous Burmese Amber." *Science* 34 (2006): 614.

Wheeler, William Morton. *Social Life Among the Insects.* London: Constable, 1923.

Wilson, Edward O. *The Insect Societies.* Cambridge, MA: Belknap Press, 1971.

BEES AND FLOWERS

Barth, Friedrich G. *Insects and Flowers: The Biology of a Partnership.* Translated by M.A. Biederman-Thorson. Princeton, NJ: Princeton University Press, 1985.

Heinrich, Bernd. *Bumblebee Economics.* Cambridge, MA: Harvard University Press, 1979.

Kevan, Peter G. "Entomology: A Celebration of Little Wonders." *Bulletin of the Entomological Society of Canada* 38 (March 2006): 4–7.

——. "Floral Colors Through the Insect Eye: What They Are and What They Mean." *Handbook of Experimental Pollination Biology.* Edited by E. Eugene Jones and R. John Little. New York: Scientific and Academic Editions, 1983: 3–30.

Laverty, T.M. "The Flower-Visiting Behaviour of Bumble Bees: Floral Complexity and Learning." *Canadian Journal of Zoology* 58 (1980): 1324–35.

Proctor, Michael, Peter Yeo, and Andrew Lack. *The Natural History of Pollination.* Portland, OR: Timber Press, 1996.

CHAPTER 2

HONEYBEE BIOLOGY

Gould, James L., and Carol Grant Gould. *The Honey Bee.* New York: Scientific American Library, 1988.

Lindauer, Martin. *Communication Among Social Bees.* Cambridge, MA: Harvard University Press, 1961.

Ratnieks, Francis L.W., and P. Kirk Visscher. "Worker Policing in the Honeybee." *Nature* 342 (1989): 796–97.

Robinson, Gene E. "Chemical Communication in Honeybees." *Science* 271 (1996): 1824–25.

——. "Regulation of Division of Labor in Insect Societies." *Annual Review of Entomology* 37 (1992): 637–65.

Seeley, Thomas D. *Honeybee Ecology: A Study of Adaptation in Social Life.* Princeton, NJ: Princeton University Press, 1985.

———. *The Wisdom of the Hive: The Social Physiology of Honey Bee Colonies.* Cambridge, MA: Harvard University Press, 1995.

Thale, Wolfgang, and Herbert Habersack, directors. "Bees: Tales From the Hive." VHS. Boston: WGBH, 2000.

von Frisch, Karl. *Bees: Their Vision, Chemical Senses, and Language.* Ithaca, NY: Cornell University Press, 1950.

———. *The Dancing Bees: An Account of the Life and Senses of the Honey Bee.* New York: Harcourt, Brace, 1953.

Whitfield, Charles W., et al. "Thrice Out of Africa: Ancient and Recent Expansions of the Honey Bee, *Apis mellifera.*" *Science* 314 (2006): 642–45.

Winston, Mark L. *The Biology of the Honey Bee.* Cambridge, MA: Harvard University Press, 1987.

———. *From Where I Sit: Essays on Bees, Beekeeping, and Science.* Ithaca, NY: Comstock, 1998.

CHAPTER 3

THE DANCING BEES

Gould, James L. "Honey Bee Recruitment: The Dance-Language Controversy." *Science* 189 (1975): 685–92.

Lindauer, Martin. "The Dance Language of Honeybees: The History of a Discovery." *Experimental Behavioral Ecology and Sociobiology: In Memoriam Karl von Frisch 1886–1982.* Edited by Bert Holldobler and Martin Lindauer. Sunderland, MA: Sinauer Associates, 1985: 129–40.

Muntz, Tania. "The Bee Battles: Karl von Frisch, Adrian Wenner and the Bee Dance Language Controversy." *Journal of the History of Biology* 38 (2005): 535–70.

Riley, J.R., U. Greggers, A.D. Smith, D.R. Reynolds, and R. Menzel. "The Flight Paths of Honeybees Recruited by the Waggle Dance." *Nature* 435 (2005): 205–207.

Robinson, Gene E. "The Dance Language of the Honey Bee: The Controversy and Its Resolution." *Cornell Plantations Quarterly* 41 (1985): 66–75.

Seeley, Thomas D., Scott Camazine, and James Sneyd. "Collective Decision-Making in Honey Bees: How Colonies Choose Among Nectar Sources." *Behavioral Ecology and Social Biology* 28 (1991): 277–90.

Seeley, Thomas D., and William F. Towne. "Tactics of Dance Choice in Honey Bees: Do Foraging Bees Compare Dances?" *Behavioral Ecology and Sociobiology* 30 (1992): 59–69.

von Frisch, Karl. *The Dancing Bees: An Account of the Life and Senses of the Honey Bee.* Translated by Dora Ilse. New York: Harcourt, Brace, 1953.

——. "Decoding the Language of the Bee." Nobel Lecture, December 12, 1973. Available online at http://www.nobelprize.org/nobel_prizes/medicine/laureates/1973/frisch-lecture.html

von Frisch, Karl, and Martin Lindauer. "Indication of Distance and Direction in the Honeybee—Round and Waggle Dance." An instructional video produced in the late 1970s. Available online at http://www.iwf.de/iwf/do/mkat/details.aspx?Signatur=C+1335

——. "The 'Language' and Orientation of the Honey Bee." *Annual Review of Entomology* 1 (1956): 45–58.

Wenner, Adrian M. *The Bee Language Controversy: An Experience in Science.* Boulder: Education Programs Improvement Corporation, 1971.

CHAPTER 4

HOUSE-HUNTING SWARMS

Beekman, Madeleine, Robert L. Fathke, and Thomas D. Seeley. "How Does an Informed Minority of Scouts Guide a Honeybee Swarm As It Flies to Its New Home?" *Animal Behaviour* 71 (2006): 161–71.

Lindauer, Martin. "House-Hunting by Honey Bee Swarms." Translated by P. Kirk Visscher, Karin Behrne, and Susanne Kuhnholz. *Zeitschrift für vergleichende Physiologie* 37 (1955): 263–324.

Seeley, Thomas D. "Born to Dance." *Natural History* 108 (June 1999): 54–57.

——. "Decision Making in Superorganisms: How Collective Wisdom Arises from the Poorly Informed Masses." *Bounded Rationality: The Adaptive Toolbox.* Edited by G. Gigerenzer and R. Selten. Cambridge, MA: MIT Press, 2001: 249–61.

——. "A Feeling and a Fondness for the Bees." *Model Systems in Behavioral Ecology: Integrating Conceptual, Theoretical, and Empircal Approaches.* Edited by Lee Alan Dugatkin. Princeton, NJ: Princeton University Press, 2001: 27–40.

——. "The Honey Bee Colony as a Superorganism." *Exploring Animal Behavior: Readings from* American Scientist. Edited by Paul W. Sherman and John Alcock. Sunderland, MA: Sinauer Associates, 2005: 304–11.

——. "Measurement of Nest Cavity Volume by the Honey Bee (*Apis mellifera*)." *Behavioral Ecology and Sociobiology* 2 (1977): 201–27.

Seeley, Thomas D., and Susanna C. Buhrman. "Group Decision Making in Swarms of Honey Bees." *Behavioral Ecology and Sociobiology* 45 (1990): 19–31.

Seeley, Thomas D., Roger A. Morse, and P. Kirk Visscher. "The Natural History of the Flight of Honey Bee Swarms." *Psyche* 86 (1979): 103–13.

Seeley, Thomas D., and P. Kirk Visscher. "Quorum Sensing During Nest-Site Selection by Honeybee Swarms." *Behavioral Ecology and Sociobiology* 56 (2004): 594–601.

Seeley, Thomas D., P. Kirk Visscher, and Kevin M. Passino. "Group Decision Making in Honey Bee Swarms." *American Scientist* 94 (2006): 220–29.

THE AMAZING MINI-BRAIN

Giurfa, Martin. "The Amazing Mini-Brain: Lessons from a Honey Bee." *Bee World* 84 (2003), no. 1: 5–18.

Giurfa, Martin, Shaowu Zhang, Arnim Jennett, Randolf Menzel, and Mandyam V. Srinivasan. "The Concepts of 'Sameness' and 'Difference' in an Insect." *Nature* 410 (2001): 930–33.

Menzel, Randolf, Gerard Leboulle, and Dorothea Eisenhardt. "Small Brains, Bright Minds." *Cell* 124 (2006): 237–39.

Menzel, Randolf, and Martin Giurfa. "Cognitive Architecture of a Mini-Brain: The Honeybee." *Trends in Cognitive Sciences* 5 (2001): 62–71.

THE HONEYBEE GENOME

Honeybee Genome Sequencing Consortium. "Insights into Social Insects from the Genome of the Honeybee *Apis mellifera.*" *Nature* 443 (2006): 931–49.

"Human Gene Number Slashed," *BBC News,* October 20, 2004. Available online at http://www.news.bbc.co.uk/1/hi/sci/tech/3760766.stm

Robinson, Gene E. "Beyond Nature and Nurture." *Science* 304 (2004): 397–99.

——. "From Society to Genes with the Honey Bee." *Exploring Animal Behavior: Readings from* American Scientist. Edited by Paul W. Sherman and John Alcock. Sunderland, MA: Sinauer Associates, 2005: 61–67.

——. "Genes and Social Behaviour." *Essays in Animal Behaviour: Celebrating 50 Years of* Animal Behaviour. Edited by Jeffrey R. Lucas and Leigh W. Simmons. London: Elsevier, 2006: 101–13.

Whitfield, Charles W., Anne-Marie Czilko, Gene E. Robinson. "Gene Expression Profiles in the Brain Predict Behavior in Individual Honey Bees." *Science* 302 (2006): 296–99.

BEE CONSERVATION

Barrionuevo, Alexei. "Honeybees Vanish, Leaving Keepers in Peril." *New York Times,* February 27, 2007. Available online at http://www.nytimes.com

Biesmeijer, J.C., et al. "Parallel Declines in Pollinators and Insect-Pollinated Plants in Britain and the Netherlands." *Science* 313 (2006): 351–54.

Buchmann, Stephen, and Gary Paul Nabhan. *The Forgotten Pollinators.* Washington, DC: Island Press, 1996.

Kevan, Peter G. "Pollinators as Bioindicators of the State of the Environment: Species, Activity and Diversity." *Agriculture, Ecosystems and Environment* 74 (1999): 373–93.

National Academy of Sciences. *Status of Pollinators in North America.* Washington, DC: National Academies Press, 2007.

"Pollination Canada Observer's Kit." Available online at http://www.seeds.ca

"The Sao Paulo Declaration on Pollinators." Recommendations of an International Workshop, Sao Paulo, 1999. Available online at http://www.rge.fmrp.usp.br/beescience/arquivospdf/workshop.pdf

Underwood, Robyn M., and Dennis vanEngelsdorp. "Colony Collapse Disorder: Have We Seen This Before?" Available online at http://www.beeculture.com

"Urban Bee Gardens: A Practical Guide to Introducing the World's Most Prolific Pollinators Into Your Garden." Available online at http://www.nature.berkeley.edu/urbanbeegardens/index.html

"Xerces Society Red List of Pollinating Insects of North America." Available online at http://www.xerces.org/Pollinator_Red_List/index.htm

PICTURE CREDITS

Except as noted below, line drawings of bees are from W. Harcourt Bath, *The Young Collector's Handbook of Ants, Bees, Dragon-Flies, Earwigs, Crickets, and Flies* (London: Swan Sonnenschein, Lowrey and Co., 1888).

INDEX

Page numbers in italics refer to captions.

Aesop, 14
Africa, 17, 24, 26
"Against Idleness and Mischief," 52
Andrenidae, 17
"Angel of Bees," 87
Antennae, 37, 41, 48, 50, 63
Apidae, 17
Apis cerana, 81
Apis genus. *See* Honeybees
Apis mellifera, 26. *See also* Honeybees;
    specific subjects
Aristotle, 49
Asia, 24, 25, 26
Asian honeybee, 81
Australia, 24, 27, 100, 106

Bee goddess, 2
Bee identification, 10, 12–13, 18.
    See also Taxonomy
Bee 107, 53–54, 56, 57, 60, 84–85
Bee orchids, 73
Beeswax, 22, 25, 27, 30, 44, 53, *94*
Biodiversity of bees, 13, 16–18

Borneo, 26
Brihadâranyaka Upanishad, 96
Brood and brood tending, 19, 21, 22, 37,
    40, 44, *44*, 49, 53, 54
Bumblebees, 17, *17*, 22–24, 25, 109
Burma, 10, 13
Busch, Wilhelm, *50, 56*
Butler, Charles, v, 33
Buzz-runners, 94–95
Brain, 49, 78, 100–101
Brazil, 24

Canada, 106
Carpenter bees, 17, 21–22, 109
Case of the Overheated Colony, 79
Cleaning dance, *81*
Co-evolution, 13
Colletidae, 17
Colony collapse disorder, 106–7
Color vision, 29, 31. *See also* Eyes and
    eyesight
Communication, 31, 53, 57, 60.
    See also Dance "language"
Communication among Social Bees, 97
Concept formation, 100–101

Cooling the hive, 48, 57, 79
Corbicula, *63*
Cretaceous period, 10, 17
Crozier, Lorna, 87

Dance "language," 61, 63, 66, *66*,
    68–70, 72–74, 77–78, 79, 81, 85,
    86, 90, 93–94
Decision making, 97–98
Developmental sequence, 53, 54, 84
Dickinson, Emily, 71
Digger bees, 19, 21, 109
Distribution of bees, 16, 26
Division of labor, 37, 44, 53, 84
DNA. *See* Genes and genetics
Drones, 22, 25, 37, 40–41, 43, 74
Dwarf honeybees, 26

Edwards, M. Ellen, *69*
Eggs and egg laying, 21, 22, 37, 40
Egypt, 30
Emerson, Ralph Waldo, 23
"End Notes for a Small History," 104–5
England, 25, 109
"Englishman's flies," 26
Europe, 17, 26, 109. *See also* France;
    Germany; Netherlands; Rhodes;
    United Kingdom
Eusociality, 25
Evolution, 10, 12–13, 74, 77.
    See also Genes and genetics
Eyes and eyesight, 29, *29*, 31, 41, 70
Extinctions, 109

Families, taxonomic, 17
Feral honeybees, 26
"Fire shitters," 24
Flies, 18
Florida, 106, 107
Flowering plants, 10, 13, 75.
    See also Pollination; Scents
Foragers and foraging, 37, 53, 57, 60,
    61, 74–75, 77–79, 85. *See also* Dance
    "language"
Fossils, 10
France, 100, 106
Frisch, Karl von, 5, 7, 9, 28–29, 31, 34, 35,
    41, 50, 51, 60, 63, 66, *66*, 67, 68–70, 72,
    73, 74, 77, 81

Gender determination, 40, 48
Gene regulation, 102
Genes and genetics, 48, 50–51, 56, 77, 101–3
Germany. *See* Frisch, Karl von; Lindauer,
    Martin; Munich
German Zoological Society, 28, 29
"Gleich und Gleich," 64
Giant honeybees, 26
Goethe, Johan Wolfang von, 64
Griffin, Donald R., 72
Guard bees, 48, 53

Habitat for bees, 16, 18–19. *See also* Nests
Halictidae, 17, 21
Harmonic radar, 73
Hive bees. *See* Inside workers
Holocaust, the, 5, 103

Honey and honey making, 3, 27, 30,
48, 49, 54, 74, 75
Honeybee genome, 101–3.
See also Genes and genetics
Honeybees, 8, 17, 25–31.
See also specific subjects
Honeycomb, 26, 30, 40, 53, *94*
Honey doctrine, 96
Honeypot, 22
Hornets, 11
House hunting. *See* Swarms
and swarming
Humble-bee. *See* Bumblebee
"Humble-Bee, The," 23

Incubation, 22
India, 25
Inside workers, 44, 56, 57, 60, 74, 78, 81,
104. *See also* Bee 107; Brood and brood
tending; Developmental sequence

Jenkinson, E.J., 4

Kapusta, Janusz, *61*
"King" bee, 38–39, 49
King Solomon, 4
Kin selection, 48

Larvae, 13, 21, 41, 44, *44*, 49, 74.
See also Brood and brood tending
Leaf-cutter bees, 17, 109
Lies, Betty, 104–5
Life span, 48

"Like and Like," 64
Lindauer, Martin, 5, 8–10, 27, 31, 34–37,
44, 50, 53, 56–57, 60, 67, 68, 77–79, 84,
86, 89, 90, 97, 100
Long-tongued bees, 17

Marking bees, 51, 53, 89
Mason bees, 17, 109
Massacre of drones, 43
Mating. *See* Reproduction
Measurement, 40, 91
Megachilidae, 17
Melittidae, 17
Memory, 63, 99
Menzel, Randolf, 100
Metamorphosis, 21, 40, 41
Mini-brain, 100, 101. *See also* Brain
Mites, 107
Munich, 5, 8, 9, 34, 74, 89

Natural selection. *See* Evolution
Nature and nurture, 103
Navigation, 69
Nazis. *See* Third Reich
Nectar, 17, 22, 43, 44, 53, 60, 61,
75, 78, 81
Nests, 19, 21, 22, 26, 89, 90–92.
Netherlands, 109
New York State, 18
Nobel Prize, 28, 73
North America, 17, 97, 109.
See also Canada; United States
Nurse bees. *See* Inside workers

Observation hives, *35, 36,* 51
Ocelli, *29*
Olfaction. *See* Scents
Orchard bees, 109
"Origins of Honeybees, The," 30
Outside workers.
    See Foragers and foraging

Paper wasps, 11
Pesticides, 107
Pliny the Elder, 3, 38–39
Pollen and pollen gathering, 13, 21,
    22, *63, 74,* 74–75
Pollination, 13, 61, 107, 109
Population declines, 106–7, 109
Propolis, *74*
Pupae, 21

Queen bees, 22, 25, 28, 37–40, 41,
    43, *44, 44,* 49, 50, *50,* 73, 74
Queen cup, *44, 49*
Queen of Sheba, 4
Queen substance, 50
Quorum formation, 94, 98

Re (Egyptian god), 30
Recruitment, 31, 57, 60.
    See also Dance "language"
Reproduction, 19, 21, 22, 40–41, 43.
    See also Swarms and swarming
Rhodes, *3*
Robinson, Gene E., 5, 103
Rösch, G.A., 51, 53, 54

Round dances, 66, *66.*
    See also Dance "language"
Royal jelly, 44

Scents, 31, 63, *63,* 66, 70, 73, 99, 100
Scout bees, 31, 60, 75, 77, 92, *92,*
    93, 94, 98
Seeley, Thomas, 5, 81, 90, 91, 97–99
Shakespeare, William, 46
Shaw, George Bernard, 83
Short-tongued bees, 17
Skeps, 26, *26*
Social bees, 21–22, 24–25. *See also*
    Bumblebees; Honeybees; Stingless bees
Solitary bees, 18–19, 21
Stenotritidae, 17
Sting, 12–13, 43
Stingless bees, 17, 24–25, 27
Sun compass, 69
Super-organism, 85–86
Swarms and swarming, 86, *89,* 89–90,
    90, 93, 93–95
Sweat bees, 17, 21, *21*
Swammerdam, Jan, *29*

Tanging, *93*
Taxonomic deficit, 18
Taxonomy, 13, 17, 25
"Telling the Bees," 11
Temperature regulation, 22, 48, 57, 74, 79
Third Reich, 5, 9, 34, 35, 103
Thoreau, Henry David, 76
Thorpe, W.H., 72

Three Habits of Highly Intelligent
    Groups, 99
Tremble dance, 81, 90, 94
Tynan, Katharine, 11

United Kingdom, 106, 109
United States, 73, 97, 106
Unloading time, 78–79, 81

Vanishing-bee syndrome, 106–7
Virgil, 3
Vision. *See* Eyes and eyesight

Waggle dances, *66*, 66–67.
    See also Dance "language"
Wasps, 11, 14
Water, 57, 78, 79
Watts, Isaac, 52
Wax. *See* Beeswax

Wenner, Adrian, 73, 74
Wilson, E.O., 1
Winter, 19, 22, 74
Worker bees, 22, 25, 28, 37, 41, 43, 49, 50,
    51, 54, 56. *See also* Bee 107; Brood and
    brood tending; Dance "language";
    Developmental sequence; Division of
    labor; Foragers and foraging; Inside
    workers
Worker policing, 49

Xerxes Society for Invertebrate
    Conservation, 109

Yeats, William Butler, 27
Yellow jackets, 11

Zoological Institute, Munich, 9, 34

DAVID SUZUKI FOUNDATION

The David Suzuki Foundation works through science and education to protect the diversity of nature and our quality of life, now and for the future.

With a goal of achieving sustainability within a generation, the Foundation collaborates with scientists, business and industry, academia, government and non-governmental organizations. We seek the best research to provide innovative solutions that will help build a clean, competitive economy that does not threaten the natural services that support all life.

The Foundation is a federally registered independent charity, which is supported with the help of over 50,000 individual donors across Canada and around the world.

We invite you to become a member. For more information on how you can support our work, please contact us:

The David Suzuki Foundation
219–2211 West 4th Avenue
Vancouver, BC
Canada v6k 4s2
www.davidsuzuki.org
contact@davidsuzuki.org
Tel: 604-732-4228
Fax: 604-732-0752

Checks can be made payable to The David Suzuki Foundation. All donations are tax-deductible.

Canadian charitable registration: (BN) 12775 6716 RR0001
U.S. charitable registration: #94-3204049